Astronomical for Calculators

FOURTH EDITION
Enlarged & Revised

Astronomical Formulæ
for Calculators
FOURTH EDITION
Enlarged & Revised

Jean Meeus
Vereniging voor Sterrenkunde Belgium

Foreword by
Roger W. Sinnott
of **Sky and Telescope** Magazine

Published by:

Willmann–Bell, Inc.
P.O. Box 35025
Richmond, Virginia 23235   ☎ (804)
United States of America   320-7016

Publishers and Booksellers

Serving Astronomers Worldwide
Since 1973

First Edition (English) published 1979 in Belgium and copyrighted as
Volume 4, Monografieen Over Astronomie En Astrofysica Uitgegeven
Door, Volkssterrenwacht Urania V.Z.W. En Vereniging Voor Sterren-
kunde V.Z.W.

First Edition (French) published in France and Copyright 1980 Société
Astronomique de France, Paris (Dépôt Légal No 1)

Library of Congress Cataloging-in-Publication Data

Meeus, Jean.
    Astronomical formulae for calculators

    Includes index.
    1. Astronomy–Data processing. 2. Astronomy–Problems,
exercises, etc. I. Title
QB51.3.E43M43 1988   523'.0212   88-27876
ISBN 0-943396-22-0

Printed in the United States of America

Foreword

Astronomy and calculating techniques have enjoyed a long and fruitful association. Over 150 years ago, so the story goes, young Charles Babbage at the University of Cambridge in England became infuriated at the number of errors he found in some astronomical tables. He muttered that he wished they had been produced by steam machinery, instead of by humans, and his friend John Herschel replied, "It is quite possible!"

This apparently offhand remark launched Babbage's lifelong obsession to design and build various calculating engines – the forerunners of today's computers. In the early 1950's, soon after the first "electronic brains" were constructed, their earliest applications included the wholesale calculation of tables and ephemerides for the astronomical almanacs.

The arrival of pocket calculators and home computers in the 1970's marked yet another advance. Logarithmic and trigono-metric tables became outmoded almost overnight. Working alone at home or on a ship at sea, anyone could now perform calculations for navigation, practical astronomy, or celestial mechanics, without access to preprinted tables or a giant computer. So novel was the approach that virtually no book existed that was devoted, purely from scratch, to the art of astronomical calculation.

The present work fills this need brilliantly. Any serious amateur or student of astronomy will find it a superb com-panion to the introductory texts of the classroom. All too often, the modern educational trend is to present important concepts vaguely. A teacher might say, for example, that nutation is a "nodding" of the Earth's axis, and then try to explain the cause. How much more meaningful it is, and how rare in any classroom book, to find in Chapter 15 the actual formulae describing this effect!

The motions of the Moon and most planets were analyzed in great detail around the turn of the present century. The results are generally locked away in dust-laden observatory annals, to be consulted only at special libraries. One can look up, for instance, E.W. Brown's lifework on the Moon, published in 1919. To find the Moon's position at a parti-cular date and time to full precision requires evaluating over 1,650 trigonometric terms. The modern texts on celestial mechanics, understandably, omit these concrete formulae and dwell on theory, leaving the practical-minded person in the dark. However, it is in fact much easier to evaluate the main terms on a pocket calculator than it is to understand the theory.

Jean Meeus has done an important service by consulting these classic treatises on the Moon and planets, selecting the main perturbations from among a multitude of smaller effects, and presenting them concisely here. In doing this, he has carefully included the so-called secular terms, which are small or nearly constant at the present time, but which grow sizable a few centuries in the past and future. Accordingly, the formulae in this book (unlike those often found elsewhere) may safely be used in serious historical investigations.

Worthwhile discoveries, too laborious a few years ago, probably await anyone who cares to spend pleasant evenings with a pocket calculator. Consider the faint star that Galileo plotted near Jupiter's satellites on December 28, 1612. The world did not know until 1980 that this "star" was actually the planet Neptune. Yet a discovery of this kind could easily be made by some reader of the present book! All the needed formulae are here – the positions of Jupiter and Neptune are deduced by the methods of Chapters 22 – 25, and the con – figuration of Jupiter's satellites, as drawn by the great Italian astronomer on that evening in 1612, are easily confirmed by the method of Chapter 36.

This book includes ingenious methods of the author's own devising, such as the simple formulae for the times of the equinoxes and solstices in Chapter 20. His many other books and articles, exhibiting astronomical calculations of all sorts, have won him high esteem in professional circles over the years. In 1981 the International Astronomical Union announced the naming of asteroid 2213 Meeus in his honor. The first edition of **Astronomical Formulae for Calculators** was eagerly snapped up in a few months. Now that it is available again in a revised and expanded form, it is sure to become a classic.

Roger W. Sinnott
Sky and Telescope Magazine

PREFACE
TO THE FIRST (BELGIAN) EDITION

With the spectacular rise of the pocket calculating machines and the even spectacular fall of their prices in recent years, these wonderful machines are now within reach of everyone. The number of amateur astronomers possessing such calculating machines nearly equals now the number of amateur astronomers themselves. The number of the latter who own a programmable calculating machine is already impressive too and always growing. It is mainly for this last category of interested persons that this book is intended.

Anyone who endeavours to make astronomical calculations has to be very familiar with the essential astronomical conceptions and rules and he must have sufficient knowledge of elementary mathematical techniques. As a matter of course he must have a perfect command of his calculating machine, knowing all possibilities it offers the competent user. However, all these necessities don't suffice. Creating useful, successful and beautiful programs requires much practice. Experience is the mother of all science. This general truth is certainly valid for the art of programming. Only by experience and practice one can learn the innumerable tricks and dodges that are so useful and often essential in a good program.

Astronomical Formulae for Calculators intends to be a guide for the amateur astronomer who wants to do calculations. Before I specify briefly the aims and contents of the book, let me outline first what it is not.

This book is not a general textbook on Astronomy. Elementary astronomical knowledge is taken for granted. For instance, no definitions are given of right ascension and declination, ecliptic, precession, magnitude, etc., but all these notions are used continually throughout the book. Only exceptionally a definition will be given. Nor is this book a textbook on mathematics or a manual for programmable pocket calculators. As I said, the reader is assumed to be able to use his machine appropriately.

1

What this book intends is to lend a helping hand to every amateur astronomer with mathematical interests and to give him much practical information, advice and examples. About forty topics in the field of calendar problems, celestial phenomena and celestial mechanics are dealt with, and also a few astronomy oriented mathematical techniques, as interpolation and linear regression. For all these cases there is an outline of the problem, its meaning and its signification. The formulae describing the problem in mathematical terms are given and treated at some lenght so as to enable the reader to use them for making his own programs. Many numerical examples are then offered to illustrate the subject and the applications of the formulae.

No programs are given. The reasons are clear. A program is useful only for one type of calculating machine. For instance, a program for a HP-67 cannot be used on a TI-59, and even not on a HP-65. Every calculator thus must learn to create his own programs. There is the added circumstance that the precise contents of a program usually depend on the specific goals of the computation, that are impossible to anticipate always by the author.

The writing of a program to solve some astronomical problem sometimes will require a study of more than one chapter of this book. For instance, in order to create a program for the calculation of the Sun's altitude for a given time of a given date at a given place, one must first convert the date and time to Julian Date (Chapter 3), then calculate the Sun's longitude for that time (Chapter 18), its right ascension (Chapter 8), the sidereal time (Chapter 7), and finally the required altitude of the Sun (Chapter 8).

It is clear that not all topics of mathematical astronomy could have been dealt with in this book. So nothing is said about orbit determination, occultations of stars by the Moon, the calculation of the longitude of the central meridian of Mars and Jupiter for a given instant, meteor astronomy, eclipsing binaries, etc. However, a hasty look on the Table of Contents convinces there is enough fascinating material in this fourth monograph on astronomy and astrophysics edited by *Urania* and *VVS*, to keep every amateur busy for years to come.

G. Bodifée

PREFACE TO THE SECOND EDITION

In this second edition several misprints and errors have been corrected. The principal change in the new edition is the addition of some material, such as a method for the calculation of the date of Easter in the Julian Calendar, the principal periodic terms in the motion of Uranus and Neptune, and a formula for the direct calculation of the equation of the center from the orbital eccentricity and the mean anomaly.

The chapter on the positions of the satellites of Jupiter has been improved, while a new chapter has been added, namely concerning the calculation of the longitude of the central meridian of Jupiter, and of the planetocentric latitude of the center of its disk.

Since the first edition of this book (November 1978), many types of microcomputers became available at reasonable prices, and many amateur astronomers have purchased such a powerful machine. With a microcomputer it's possible to write much more sophisticated and longer programs, and to perform much more accurate calculations, than with a pocket calculator. For instance, the position of the Sun can be calculated with full accuracy, the times of a solar eclipse can be found with a precision better than one second, and so on.

Nevertheless, the present book remains intended principally for the owner of a pocket calculator. Detailed methods and formulae, designed for programs on a microcomputer, are wholly outside the scope of this book : they would require a much larger work than this one.

Jean Meeus
April 1982

PREFACE TO THE FOURTH EDITION

The only change in this edition is the addition of a chapter on the calculation of the heliocentric position of Pluto.

Jean Meeus
October 1988

CONTENTS

SOME SYMBOLS AND ABBREVIATIONS

e Eccentricity (of an orbit)
h Altitude above the horizon
r Radius vector, or distance of a body to the Sun, in AU
v True anomaly

A Azimuth
H Hour angle
M Mean anomaly
R Distance from Sun to Earth, in AU

α Right ascension
δ Declination
ϵ Obliquity of the ecliptic
θ Sidereal time
θ_0 Sidereal time at Greenwich
π Parallax
ϕ Geographical latitude
ϕ' Geocentric latitude

Δ Distance to the Earth, in AU
ΔT Difference ET – UT
$\Delta\epsilon$ Nutation in obliquity
$\Delta\psi$ Nutation in longitude
$\odot$ Geocentric longitude of the Sun

AU Astronomical unit
ET Ephemeris Time
UT Universal Time
JD Julian Day

INT Integer part of

A.E. Astronomical Ephemeris

IAUC International Astronomical Union Circular

7

Following a general astronomical practice (see for instance the *A.E.*), small superior symbols are placed immediately above the decimal point, not after the last decimal. For instance, $28\overset{\circ}{.}5793$ means 28.5793 degrees.

Moreover, note carefully the difference between hours with decimals, and hours – minutes – seconds. For example, $1\overset{h}{.}30$ is *not* 1 hour and 30 minutes, but 1.30 hours, that is 1 hour and 30 hundredths of an hour, or $1^{h}18^{m}$.

The author wishes to express his gratitude to Mr. G. Bodifée, for his valuable advice and assistance.

1

Hints and Tips

To explain how to calculate or to program on a calculating machine is out of the scope of this book. The reader should, instead, study carefully his instructions manual. However, even then writing good programs cannot be learned in the lapse of one day. It is an art which can be acquired only progressively. Only by practice one can learn to write better and shorter programs.

In this first Chapter, we will give some practical hints and tips, which may be of general interest.

Accuracy

The accuracy of a calculation depends on its aims. If one only wants to know whether an occultation by the Moon will be visible in some countries, an accuracy of 100 kilometers in the northern or southern limit of the region of visibility is probably sufficient ; however, if one wants to organize an expedition to observe a grazing occultation by the Moon, the limit has to be calculated with an accuracy better than 1 kilometer.

If one wants to calculate the position of a planet with the goal of obtaining the moments of rise or setting, an accuracy of 0.01 degree is sufficient. But if the position of the planet is needed to calculate the occultation of a star by the planet, an accuracy of better than $1''$ will be necessary because of the small size of the planet's disk.

To obtain a better accuracy it is sometimes necessary to use another method of calculation, not just to keep more decimals in the result of an approximate calculation. For example, if one has to know the position of Mars with an accuracy of 0.1 degree, it suffices to use an unperturbed elliptical orbit (Keplerian motion) although secular perturbations of the orbit are to be taken into account eventually. However, if the position of Mars is to be known with a precision of $10''$ or better, perturbations due to the other planets have to be calculated and the program will be a much longer one.

So the calculator, who knows his formulae and the desired accuracy in a given problem, must himself considerate which terms, if any, may be omitted in order to keep the program handsome and as short as possible. For instance, the geometric mean longitude of the Sun, referred to the mean equinox of the date, is given by

$$L = 279°41'48\rlap{.}''04 + 129\ 602\ 768\rlap{.}''13\ T + 1\rlap{.}''089\ T^2$$

where T is in Julian centuries of 36525 ephemeris days from the epoch 1900 January 0.5 ET. In this expression the last term (secular acceleration of the Sun) is smaller than $1''$ if $|T| < 0.96$, that is between the years 1804 and 1996. If an accuracy of $1''$ is sufficient, the term in T^2 may thus be dropped for any instant in that period. But for the year -100 we have $T = -20$, so that the last term becomes $436''$, which is larger than 0.1 degree.

Rounding

Rounding should be made where it is necessary. Do not retain meaningless decimals in your result. Some "feeling" and sufficient astronomical knowledge are necessary here. For instance, it would be completely irrelevant to give the illuminated fraction of the Moon's disk with an accuracy of 0.000 000 001.

If one calculates by hand and not with a program, the rounding should be performed *after* the whole calculation has been made.

Example : Calculate 1.4 + 1.4 to the nearest integer. If we first round the given numbers, we obtain 1 + 1 = 2. In fact, 1.4 + 1.4 = 2.8, which is to be rounded to 3.

Rounding should be made to the nearest value. For instance, 15.88 is to be rounded to 15.9 or to 16, not to 15. However, calendar dates and years are exceptions. For example, March 15.88 denotes an instant belonging to March 15 ; thus, if we read that an event occurs on March 15.88, it takes place on March 15, not on March 16. Similarly, 1977.69 denotes an event occurring in the year 1977, not 1978.

Trigonometric functions of large angles

Large angles frequently appear in astronomical calculations. In Example 18.a we find that on 1978 November 12.0 the Sun's mean longitude is 28670.77554 degrees. Even larger angles are found for rapidly moving objects, such as the Moon or the bright satellites of Jupiter.

According to the type of the machine, it may be necessary or desirable to reduce the angles to the range 0 – 360 degrees. Some

calculating machines give incorrect values for the trigonometric
functions of large angles. For instance,

the HP-55 gives   sin 360000030° = 0.499 481 3556
TI-52   0.499 998 1862
Casio fx 2200   Error

while the HP-67 gives the correct value 0.500 000 0000.

Angle modes

The calculating machines do not calculate directly the trigonometric functions of an angle which is given in degrees, minutes
and seconds. Before performing the trigonometric function, the
angle should be converted to degrees and *decimals*. Thus, to calculate the cosine of 23°26'49", first convert the angle to
23.44694444 degrees, and *then* press the key COS.

Similarly, angles should be converted from degrees, minutes and
seconds to degrees and decimals, before they can be interpolated.
For instance, it is impossible to apply an interpolation formula
directly to the values

$$5°03'45''$$
$$5°34'22''$$
$$6°17'09''$$

Right ascensions

Right ascensions are generally expressed in hours, minutes and
seconds. If the trigonometric function of a right ascension must
be calculated, it is thus necessary to convert that right ascension to degrees. Remember that one hour corresponds to 15 degrees.

Example 1.a :   Calculate tan α, where α = $9^h14^m55\overset{s}{.}8$.

We first convert α to hours and decimals :

$$9^h14^m55\overset{s}{.}8 = 9.248\ 833\ 333 \text{ hours.}$$

Then, by multiplying by 15,

$$α = 138\overset{\circ}{.}73250$$

whence tan α = −0.877 517.

The correct quadrant

When the sine, the cosine or the tangent of an angle is known, the angle itself can be obtained by pressing the corresponding key : arc sin, arc cos, or arc tan, sometimes written as $\sin^{-1}$, $\cos^{-1}$, $\tan^{-1}$. — The latter are, in fact, incorrect designations, for x^{-1} is the same as $1/x$. But $\cos^{-1} x$ is (incorrectly) used to designate the inverse function, and *not* $1/\cos x$.

In this case, on most pocket calculating machines, arc sin and arc tan give an angle lying between −90 and +90 degrees, while arc cos gives a value between 0 and +180 degrees.

In some cases, the result obtained in this way may not be in the correct quadrant. Each problem must be examined separately. For instance, formulae (8.4) and (25.15) give the sine of the declination. The instruction arc sin will then give the declination always in the correct quadrant, because all declinations lie between −90 and +90 degrees.

This is also the case for the angular separation whose cosine is given by formula (9.1). Indeed, any angular separation lies between the values 0° and 180°, and this is precisely what the operation arc cos gives.

When the tangent of an angle is given, for example by means of formulae (8.1), (8.3) and (18.3), the angle may be obtained directly in the correct quadrant by using a trick : the rectangular/polar transformation applied to the numerator and the denominator of the fraction in the right-hand member of the formula, as explained in Chapter 8 and at some other places in this book.

Powers of time

Some quantities are calculated by means of a formula containing powers of the time (T, T^2, T^3, ...). It is important to note that such polynomial expressions are valid only for values of T which are not too large. For instance, the formula

$$e = 0.016\ 751\ 04 - 0.000\ 0418\,T - 0.000\ 000\ 126\ T^2$$

given in Chapter 18 for the eccentricity of the Earth's orbit, is valid only for several centuries before and after the year 1900, and *not* for millions of years ! For instance, for $T = 1000$, the above-mentioned formula gives $e = -0.151 < 0$, an absurd result.

The same is true for instance for formula (18.4), which would give the completely invalid results $\varepsilon = 0°$ for $T = -383$, and $\varepsilon = 90°$ for $T = +527$.

One should further carefully note the difference between periodic terms, which remain small throughout the centuries, and secular terms (terms in T^2, T^3, ...) which rapidly increase with time.

In formula (32.1), for instance, the last term is a periodic one which always lies between −0.00033 and +0.00033. On the other hand, the term +0.000 1178 T^2, which is very small when T is very small, becomes increasingly important for larger values of $|T|$. For $T = \pm10$, that term takes the value +0.01178, which is large in comparison to the above-mentioned periodic term. Thus, for large values of T it is meaningless to take into account small periodic terms if secular terms are dropped.

To shorten a program

To make a program as short as possible is not always an art for art's sake, but sometimes a necessity as long as the memory capacities of the calculating machine have their limits.

There exist many tricks to make programs shorter, even for simple calculations. For instance, if one wants to calculate the polynomial

$$Ax^4 + Bx^3 + Cx^2 + Dx + E$$

with A, B, C, D and E constants, and x a variable. Now, one may program the machine directly to calculate this polynomial term after term and adding all terms, so that for each given x the machine obtains the value of the polynomial. However, instead of calculating all the powers of x, it appears to be wiser to write the polynomial as follows :

$$\left[\left((Ax + B)x + C\right)x + D\right]x + E$$

In this expression all power functions have disappeared and only additions and multiplications are to be performed. The program will be shorter now. If we use for instance a HP-67 machine and store the constants A to E in the registers 1 to 5, the programs for the calculation will in each case be as follows.

First version	Second version
STO A	STO A
4	RCL 1
y^x	×
RCL 1	RCL 2
×	+
RCL A	RCL A
3	×
y^x	RCL 3
RCL 2	+
×	RCL A
+	×
RCL A	RCL 4
x^2	+
RCL 3	RCL A
×	×
+	RCL 5
RCL A	+
RCL 4	
×	
+	
RCL 5	
+	

Thus, by using this simple trick, one has saved five steps, a gain of 23 % in this short program !

2

INTERPOLATION

The astronomical almanacs or other publications contain numerical tables giving some quantities y for *equidistant* values of an argument x. For example, y is the right ascension of the Sun, and the values x are the different days of the year at 0^h ET.

Interpolation is the process of finding values for instants, quantities, etc., intermediate to those given in a table.

In this Chapter we will consider two cases : interpolation from three or from five tabular values. In both cases we will also show how an extremum or a zero of the function can be found. The case of only two tabular values will not be considered here, for in that case the interpolation can but be linear, and this will give no difficulty at all.

Three tabular values

Three tabular values y_1, y_2, y_3 of the function y are given, corresponding to the values x_1, x_2, x_3 of the argument x. Let us form the table of differences

$$
\begin{array}{lll}
x_1 & y_1 & \\
 & & a \\
x_2 & y_2 & \quad c \\
 & & b \\
x_3 & y_3 & \\
\end{array}
\qquad (2.1)
$$

where $a = y_2 - y_1$ and $b = y_3 - y_2$ are called the *first differences*. The *second* difference c is equal to $b - a$, that is

$$c = y_1 + y_3 - 2y_2$$

Generally, the differences of the successive orders are gradually smaller. Interpolation from three tabular values is permitted when the second differences are almost constant in that part of the table, that is when the third differences are almost zero. Let us consider, for instance, the distance of Mars to the Earth from 4 to 8 August 1969, at 0^h ET. The values are given in astronomical units, and the differences are in units of the sixth decimal :

August 4	0.659441			
		+5090		
5	0.664531		+30	
		+5120		−1
6	0.669651		+29	
		+5149		0
7	0.674800		+29	
		+5178		
8	0.679978			

Since the third differences are almost zero, we may interpolate from only three tabular values.

The central value y_2 must be choosen in such a manner that it is that value of y that is closest to the required value.

For example, if from the table above we must deduce the value of the function for August 6 at 22^h14^m, then y_2 is the value for August 7.00. In that case, we should consider the tabular values for August 6, 7 and 8, namely the table

$$
\begin{array}{lll}
\text{August 6} & y_1 = 0.669651 & \\
7 & y_2 = 0.674800 & \quad\quad (2.2)\\
8 & y_3 = 0.679978 &
\end{array}
$$

and the differences are

$$
\begin{array}{ll}
a = +0.005149 & \\
b = +0.005178 & c = +0.000029
\end{array}
$$

Let n be the interpolation interval. That is, if the value y of the function is required for the value x of the argument, we have $n = x - x_2$ in units of the tabular interval. The number n is positive if $x > x_2$, that is for a value "later" than x_2, or from x_2 towards the bottom of the table. If x precedes x_2, then $n < 0$.

If y_2 has been correctly choosen, then n will be between −0.5 and +0.5, although the following formulae will also give correct results for all values of n between −1 and +1.

The interpolation formula is

$$y = y_2 + \frac{n}{2}(a + b + nc) \qquad (2.3)$$

Example 2.a : From the table (2.2), calculate the distance of Mars
to the Earth on 1969 August 7 at 4^h21^m ET.

We have 4^h21^m = 4.35 hours and, since the tabular interval is
1 day or 24 hours, we have n = 4.35/24 = 0.18125.
Formula (2.3) then gives y = 0.675 736, the required value.

If the tabulated function reaches an *extremum* (that is, a maximum or a minimum value), this extremum can be found as follows.
Let us again form the difference table (2.1) for the appropriate
part of the ephemeris. The extreme value of the function then is

$$y_m = y_2 - \frac{(a + b)^2}{8c}$$

and the corresponding value of the argument x is given by

$$n_m = -\frac{a + b}{2c}$$

in units of the tabular interval, and again measured from the central value x_2.

Example 2.b : Calculate the time of passage of Mars through the
perihelion of its orbit in January 1966, and the value of Mars'
radius vector at that instant.

From the *Astronomical Ephemeris* we take the following values for
the distance Sun - Mars :

1966 January 11.0	1.381 701
15.0	1.381 502
19.0	1.381 535

The differences are a = −0.000199 c = +0.000232
b = +0.000033

from which we deduce

$$y_m = 1.381\ 487 \qquad \text{and} \qquad n_m = +0.35776$$

The least distance from Mars to the Sun was thus 1.381 487 AU. The corresponding time is found by multiplying 4 days (the tabular interval) by +0.35776. This gives 1.43104 day, or 1 day and 10 hours later than the central time, or 1966 January 16 at 10^h.

The value of the argument x for which the function y becomes zero can be found by again forming the difference table (2.1) for the appropriate part of the ephemeris. The interpolation interval corresponding to a zero of the function is then given by

$$n_o = \frac{-2y_2}{a + b + cn_o} \qquad (2.4)$$

Equation (2.4) can be solved by first putting $n_o = 0$ in the second member. Then the formula gives an approximate value for n_o. This value is then used to calculate the right hand side again, which gives a still better value for n_o. This process, called *iteration* (Latin : *iterare* = to repeat), can be continued until the value found for n_o does not longer vary, to the precision of the calculating machine.

Example 2.c : The *A.E.* gives the following values for the declination of Mercury :

1973 February 26.0	$-0°$	28' 13".4
27.0	+0	06 46.3
28.0	+0	38 23.2

Calculate when the planet's declination was zero.

We firstly convert the tabulated values into seconds of a degree, and then form the differences :

$y_1 = -1693.4$

$\qquad a = +2099.7$

$y_2 = + 406.3 \qquad\qquad\qquad c = -202.8$

$\qquad b = +1896.9$

$y_3 = +2303.2$

Formula (2.4) then becomes

$$n_o = \frac{-812.6}{+3996.6 - 202.8\, n_o}$$

Putting $n_o = 0$ in the second member, we find $n_o = -0.20332$. Repeating the calculation, we find successively -0.20125 and -0.20127. Thus $n_o = -0.20127$ and therefore, the tabular interval being one day, Mercury crossed the celestial equator on

$$1973 \text{ February } 27.0 - 0.20127 = \text{February } 26.79873$$
$$= \text{February } 26 \text{ at } 19^h 10^m \text{ ET.}$$

Five tabular values

When the third differences may not be neglected, more than three tabular values must be used. Taking five consecutive tabular values, y_1 to y_5 , we form, as before, the difference table

$$
\begin{array}{ccccc}
y_1 & & & & \\
& A & & & \\
y_2 & & E & & \\
& B & & H & \\
n \Big\downarrow \quad y_3 & & F & & K \\
& C & & J & \\
y_4 & & G & & \\
& D & & & \\
y_5 & & & &
\end{array}
$$

where $A = y_2 - y_1$, $H = F - E$, etc. If n is the interpolation interval, measured from the central value y_3 towards y_4 in units of the tabular interval, we have the interpolation formula

$$y = y_3 + \frac{n}{2}(B + C) + \frac{n^2}{2} F + \frac{n(n^2 - 1)}{12}(H + J) + \frac{n^2(n^2 - 1)}{24} K$$

which may also be written   (2.5)

$$y = y_3 + n\left(\frac{B + C}{2} - \frac{H + J}{12}\right) + n^2\left(\frac{F}{2} - \frac{K}{24}\right) + n^3\left(\frac{H + J}{12}\right) + n^4\left(\frac{K}{24}\right)$$

Example 2.d : The *A.E.* gives the following values for the Moon's horizontal parallax :

1979 December	9.0	54'45".5099
	9.5	54 34.4060
	10.0	54 25.6303
	10.5	54 19.3253
	11.0	54 15.5940

The differences (in ") are

$A = -11.1039$

$\qquad E = +2.3282$

$B = -\ 8.7757$

$\qquad\qquad\qquad H = +0.1425$

$\qquad F = +2.4707$

$\qquad\qquad\qquad\qquad\qquad K = -0.0395$

$C = -\ 6.3050$

$\qquad\qquad\qquad J = +0.1030$

$\qquad G = +2.5737$

$D = -\ 3.7313$

We see that the third differences may not be neglected, unless an accuracy of about 0".1 is sufficient.

Let us now calculate the Moon's parallax on December 10 at 3^h20^m ET. The tabular interval being 12 hours, we find

$$n = +0.277\ 7778 \ .$$

Formula (2.5) then gives

$$y = 54'25".6303 - 2".0043 = 54'23".6260.$$

The interpolation interval n_m corresponding to an extremum of the function may be obtained by solving the equation

$$n_m = \frac{6B + 6C - H - J + 3n_m^2\,(H + J) + 2n_m^3 K}{K - 12F} \qquad (2.6)$$

As before, this may be performed by iteration, firstly putting $n_m = 0$ in the second member. Once n_m is found, the corresponding value of the function can be calculated by means of formula (2.5).

Finally, the interpolation interval n_o corresponding to a zero of the function may be found from

$$n_o = \frac{-24y_3 + n_o^2\,(K - 12F) - 2n_o^3\,(H + J) - n_o^4 K}{2\,(6B + 6C - H - J)} \qquad (2.7)$$

where, again, n_o can be found by iteration, starting from putting $n_o = 0$ in the second member.

Note that the quantities $(6B + 6C - H - J)$, $(K - 12F)$, and $(H + J)$ appear in both formulae (2.6) and (2.7). Consequently, it may be useful to calculate these quantities in a subroutine which will be used in both cases.

Exercise. - From the following values of the heliocentric latitude of Mercury, find the instant when the latitude is zero, by using formula (2.7).

1979 May 25.0 ET	$-1°16'00''5$
26.0	$-0\ 33\ 01.3$
27.0	$+0\ 11\ 12.0$
28.0	$+0\ 56\ 03.3$
29.0	$+1\ 40\ 52.2$

Answer : Mercury reaches the ascending node of its orbit for $n_o = -0.251\ 360$, that is on 1979 May 26 at 17^h58^m ET.

Important remarks

1. Interpolation cannot be performed on complex quantities directly. These quantities should be converted, in advance, into a single suitable unit. For instance, angles expressed in degrees, minutes and seconds should be expressed either in degrees and decimals, or in seconds.

Thus, for instance, $12°44'03''7$ should be written either as $12°73436$, or as $45843''7$.

2. *Interpolating times and right ascensions.* - We draw attention on the fact that the time and the right ascension jump to zero when the value of 24 hours is reached. This should be taken into account when interpolation is performed on tabulated values. Suppose, for example, that we wish to calculate the right ascension of Mercury for the instant 1979 April 16.2743 ET, using three tabulated values. We find in the *A.E.* :

1979 April 15.0	$\alpha = 23^h56^m09^s20$
16.0	$23\ 58\ 46.63$
17.0	$0\ 01\ 36.80$

Not only is it necessary to convert these values to hours and decimals, but the last value should be written as $24^h\ 01^m\ 36^s80$, otherwise the machine will consider that, from April 16.0 to 17.0, the value of α *decreases* from 23^h58^m... to 0^h01^m....

We find a similar situation in some other cases. For instance, here is the longitude of the central meridian of the Sun for a few dates :

1979 December	25.0	37°39
	26.0	24.22
	27.0	11.05
	28.0	357.88

It is evident that the variation is −13.17 degrees per day. Thus, one should *not* interpolate directly between 11.05 and 357.88. Either the first value should be written as 371.05, or the second one should be considered as being −2.12.

3

JULIAN DAY AND CALENDAR DATE

In this Chapter we will give a method for converting a date in the Julian or Gregorian calendars into the corresponding Julian Day number (JD), and vice versa.

General remarks

The Julian Day begins at Greenwich mean *noon*, that is at 12^h Universal Time (or 12^h Ephemeris Time, and in that case the expression Julian Ephemeris Day is generally used). For example, 1977 April 26.4 = JD 2443 259.9.

In the methods described below, the Gregorian calendar reform is taken into account. Thus, the day following 1582 October 4 is 1582 October 15.

The "B.C." years are counted astronomically. Thus, the year before the year +1 is the year zero, and the year preceding the latter is the year −1.

We will indicate by INT (x) the integer part of x, that is the integer which precedes its decimal point. For example :

INT $(7/4) = 1$ INT $(5.9999) = 5$
INT $(8/4) = 2$ INT $(-4.98) = -4$
INT $(5.02) = 5$

Calculation of the JD

A date may be entered in the machine as consecutive numbers, for instance the year first, then the month number, and finally the day with decimals. Thus, 1976 August 22.09 can be entered by entering successively the numbers 1976, 8 and 22.09.

However, it may be more interesting to enter a date as one single number, namely as *YYYY.MMDDdd*, where *YYYY* is the year, *MM* the month, and *DDdd* the day of the month with decimals. In that case, the month number should always be written as a two-digit number, and a decimal point must separate $YYYY$ from MM. For example, 1976 August 22.09 should then be entered as 1976.082209. The program must then start with a procedure separating the numbers $YYYY$, MM and $DD.dd$ and storing them in suitable registers. For example, for 1976 August 22.09, the number 1976.082209 is given to the machine, which stores $YYYY = 1976$ in one register, $MM = 8$ in a second one, and $DD.dd$ = 22.09 in a third register.

In what follows, we will suppose that this separation has been performed.

If MM is greater than 2, take

$$y = YYYY \qquad \text{and} \qquad m = MM \; ;$$

if $MM = 1$ or 2, take

$$y = YYYY - 1 \qquad \text{and} \qquad m = MM + 12.$$

If the number $YYYY.MMDDdd$ is equal or larger than 1582.1015 (that is, in the Gregorian calendar), calculate

$$A = \text{INT}\left(\frac{y}{100}\right) \qquad\qquad B = 2 - A + \text{INT}\left(\frac{A}{4}\right)$$

If $YYYY.MMDDdd < 1582.1015$, it is not necessary to calculate A and B.

The required Julian Day is then

$$JD = \text{INT}\,(365.25\,y) + \text{INT}\,\bigl(30.6001\,(m+1)\bigr) + DD.dd + 1720\,994.5$$

$$(3.1)$$

and, to this result, add the quantity B *if* the date is in the Gregorian calendar.

Example 3.a : Calculate the JD corresponding to 1957 October 4.81, the time of launch of Sputnik 1.

Because $MM = 10$ is greater than 2, we have $y = 1957$ and $m = 10$.

Because 1957.100481 > 1582.1015, the date is in the Gregorian calendar, and we calculate

$$A = \text{INT}\left(\frac{1957}{100}\right) = \text{INT}\,(19.57) = 19$$

$$B = 2 - 19 + \text{INT}\left(\frac{19}{4}\right) = 2 - 19 + 4 = -13$$

$$JD = INT(365.25 \times 1957) + INT(30.6001 \times 11)$$
$$+ 4.81 + 1720\ 994.5 - 13$$

$$JD = 2436\ 116.31$$

Example 3.b : Calculate the JD corresponding to January 27 at 12^h of the year 333.

Because $MM = 1$, we have

$$y = 333 - 1 = 332 \qquad \text{and} \qquad m = 1 + 12 = 13.$$

The number $YYYY.MMDDdd = 333.01275$ being less than 1582.1015, the date is in the Julian calendar, and the quantities A and B are not needed.

$$JD = INT(365.25 \times 332) + INT(30.6001 \times 14) + 27.5 + 1720\ 994.5$$

$$JD = 1842\ 713.0$$

Note. – Your program will not work for negative years. One reason is that, if you enter the date as $YYYY.MMDDdd$ preceded by a minus sign, the machine will read MM and $DD.dd$ as negative numbers. For example, if May 28.63 of the year −584 is entered as −584.052863, the machine will correctly deduce $YYYY = -584$, but will find $MM = -5$ and $DD.dd = -28.63$ instead of the correct values +5 and +28.63.

You may make your program valid for negative years by correcting it as follows.

1. After $YYYY$ has been deduced (with proper sign) from the number $YYYY.MMDDdd$, take the absolute value of $.MMDDdd$ before calculating MM and $DD.dd$;

2. If $y < 0$, replace, in formula (3.1),
$$INT(365.25\,y) \qquad \text{by} \qquad INT(365.25\,y - 0.75).$$

As an exercise, try your corrected program on −584 May 28.63. The result should be JD = 1507 900.13. But check whether your program is still valid for positive years !

Calculation of the Calendar Date from the JD

The following method is valid for positive as well as for negative years, but not for negative Julian Day numbers.

Add 0.5 to the JD, and let Z be the integer part, and F the fractional (decimal) part of the result.

If $Z < 2299\,161$, take $A = Z$.

If Z is equal to or larger than $2299\,161$, calculate

$$\alpha = \text{INT} \left(\frac{Z - 1867\,216.25}{36524.25} \right)$$

$$A = Z + 1 + \alpha - \text{INT} \left(\frac{\alpha}{4} \right)$$

Then calculate

$$B = A + 1524$$

$$C = \text{INT} \left(\frac{B - 122.1}{365.25} \right)$$

$$D = \text{INT} (365.25\,C)$$

$$E = \text{INT} \left(\frac{B - D}{30.6001} \right)$$

The day of the month (with decimals) is then

$$B - D - \text{INT} (30.6001\,E) + F$$

The month number m is
$$
\begin{array}{lll}
E - 1 & \text{if} & E < 13.5 \\
E - 13 & \text{if} & E > 13.5
\end{array}
$$

The year is
$$
\begin{array}{lll}
C - 4716 & \text{if} & m > 2.5 \\
C - 4715 & \text{if} & m < 2.5
\end{array}
$$

Example 3.c : Calculate the calendar date corresponding to JD 2436 116.31.

$$2436\,116.31 + 0.5 = 2436\,116.81,$$

thus $Z = 2436\,116$ and $F = 0.81$

Because $Z > 2299\,161$, we have

$$\alpha = \text{INT} \left(\frac{2436\,116 - 1867\,216.25}{36524.25} \right) = 15$$

26

$$A = 2436\ 116 + 1 + 15 - \text{INT}\left(\frac{15}{4}\right) = 2436\ 129$$

Then we find

$B = 2437\ 653$, $C = 6673$, $D = 2437\ 313$,
$E = 11$,

day of month = 4.81
month $m = E - 1 = 10$ (because $E < 13.5$)
year $= C - 4716 = 1957$ (because $m > 2.5$)

Thus, the required date is 1957 October 4.81

Exercises :   Calculate the calendar dates corresponding to
JD = 1842 713.0   and to   JD = 1507 900.13 .

(Answers :   333 January 27.5 and -584 May 28.63)

Time interval in days

     The number of days between two calendar dates can be found by
calculating the difference between their corresponding Julian Days.

Example 3.d :   The periodic comet Halley passed through perihelion
on 1835 November 16 and on 1910 April 20. What is the time interval between these two passages ?

1835 November 16.0   corresponds to   JD 2391 598.5
1910 April 20.0   corresponds to   JD 2418 781.5

The difference is 27 183 days.

Exercise :   Find the date exactly 10 000 days after 1954 June 30.
(Answer :   1981 November 15)

Day of Week

     The day of the week corresponding to a given date can be obtai-
ned as follows. Compute the JD for that date at 0^h, add 1.5, and
divide the result by 7. The remainder of this division will indicate the weekday, as follows : if the remainder is 0, it is a Sunday, 1 a Monday, 2 a Tuesday, 3 a Wednesday, 4 a Thursday, 5 a
Friday, 6 a Saturday.

Example 3.e : Find the weekday of 1954 June 30.

1954 June 30.0 corresponds to JD 2434 923.5
2434 923.5 + 1.5 = 2434 925
The remainder after division by 7 is 3. Thus it was a Wednesday.

Day of the Year

The number N of a day in the year can be computed as follows.

For common years :

$$N = \text{INT} \left(\frac{275\,M}{9} \right) - 2\,\text{INT} \left(\frac{M + 9}{12} \right) + D - 30$$

For leap (bissextile) years :

$$N = \text{INT} \left(\frac{275\,M}{9} \right) - \text{INT} \left(\frac{M + 9}{12} \right) + D - 30$$

where M is the month number, and D is the day of the month.

N takes integer values, from 1 on January 1 to 365 (or 366 in leap years) on December 31.

Example 3.f : 1978 November 14.

Common year, $M = 11$, $D = 14$.
One finds $N = 318$.

Example 3.g : 1980 April 22.

Leap year, $M = 4$, $D = 22$.
One finds $N = 113$.

Let us now consider the reverse problem : the day number N in the year is known, and we wish to find the corresponding date, namely the month number M and the day D of that month. This can be performed as follows.

Let $A = 1889$ in the case of a common year, $A = 1523$ in the case of a leap year. Then calculate

$$B = \text{INT} \left(\frac{N + A - 122.1}{365.25} \right)$$

28

$C = N + A - \text{INT}(365.25\,B)$

$E = \text{INT}(C/30.6001)$

$M = E - 1$ if $E < 13.5,$ $M = E - 13$ if $E > 13.5$

$D = C - \text{INT}(30.6001\,E)$

Example 3.h :   Common year, $N = 222$.

One finds successively :
$A = 1889,$ $B = 5,$ $C = 285,$ $E = 9,$ $M = 9 - 1 = 8,$ $D = 10$.
Hence, the date is August 10.

Important note on the Integer Part, INT

The microcomputers and several pocket calculators have the INT ("Integer Part") function. It is important to note that these functions differ with negative numbers.

On the microcomputers, the INT function is defined as follows : INT (x) is the greatest integer less than or equal to x. In that case, we have for instance INT $(-7.83) = -8$, because -7 is indeed larger than -7.83.

On the pocket calculators, INT is really the integer part of the *written* number, that is the part of the number that precedes the decimal point ; for instance, INT $(-7.83) = -7$.

The HP-85 microcomputer has not only the INT function, but also an IP ("integer part") function which is identical to the INT one of the pocket calculators. Hence, on the HP-85, we have

INT $(-7.83) = -8$ and   IP $(-7.83) = -7$.

Hence, take care when using the INT function ; it depends on your calculating machine ! The INT function used on the preceding pages is identical to the INT function of the pocket calculators.

4

DATE OF EASTER

The method described below has been given by Spencer Jones in his book *General Astronomy* (pages 73-74 of the edition of 1922). It has been published again in the *Journal of the British Astronomical Association*, Vol. 88, page 91 (December 1977) where it is said that it was devised in 1876 and appeared in Butcher's *Ecclesiastical Calendar*.

Unlike the formula given by Gauss, this method has no exception and is valid for all years in the *Gregorian calendar*, that is from the year 1583 on. The procedure for determining the date of Easter is as follows :

Divide	by	Quotient	Remainder
the year x	19	–	a
the year x	100	b	c
b	4	d	e
$b + 8$	25	f	–
$b - f + 1$	3	g	–
$19a + b - d - g + 15$	30	–	h
c	4	i	k
$32 + 2e + 2i - h - k$	7	–	l
$a + 11h + 22l$	451	m	–
$h + l - 7m + 114$	31	n	p

Then n = number of the month (3 = March, 4 = April),
$p + 1$ = day of that month upon which Easter Sunday falls.

Try to have your result displayed in one of the following formats :

DD.M (day.month), for instance 26.3 = 26 March ;

M.DD (month.day), for instance 3.26 = March 26 ;

YYYY.MMDD (year.month day), for instance 1978.0326 = 1978 March 26.

The month and the day of the month may also be displayed successively as integer numbers, but the formats above have the advantage that the complete date is read at a glance.

The calculation of the remainder of a division must be programmed carefully. Suppose that the remainder of the division of 34 by 30 should be found. On the HP-67 machine, we find

$$34/30 = 1.133\ 333\ 333$$

the fractional part of which is 0.133 333 333. When multiplied by 30, this gives 3.999 999 990. This result differs from 4, the correct value, and may give a wrong date for Easter at the end of the calculation.

On the HP-67, the correct value of the remainder may be found by using the instructions

DSP 0

f RND

On other machines, it may be necessary to use another trick.

If you have enough program steps, you might add some tests at the beginning of your program. For instance, write your program in such a way that *"Error"* appears if the year is not an integer number.

Try your program on the following years :

1978 → March 26	1954 → April 18
1979 → April 15	2000 → April 23
1980 → April 6	1983.6 → Error

The extreme dates of Easter are March 22 (as in 1818 and 2285) and April 25 (as in 1886, 1943, 2038).

Julian Easter

In the Julian Calendar, the date of Easter can be found by means of the following method :

Divide	by	Quotient	Remainder
the year x	4	–	a
the year x	7	–	b
the year x	19	–	c
$19c + 15$	30	–	d
$2a + 4b - d + 34$	7	–	e
$d + e + 114$	31	f	g

Then f = number of the month (3 = March, 4 = April),
   $g + 1$ = day of that month upon which Easter Sunday falls.

The date of the *Julian* Easter has a periodicity of 532 years. For instance, we find April 12 for the years 179, 711 and 1243.

5

EPHEMERIS TIME AND UNIVERSAL TIME

The Ephemeris Time (ET) is a uniform time based on the planetary motions. The Universal Time (UT), necessary for civil life, is based on the rotation of the Earth.

Because the Earth's rotation is slowing down — and, moreover, with unpredictable irregularities — UT is not a uniform time. Since the astronomers need a uniform time, they use ET for the calculation of their accurate ephemerides.

The exact value of the difference ΔT = ET − UT can be deduced only from observations. Table 5.A gives the value of ΔT for some years.

Table 5.A
Value of ΔT in minutes of time

Year	ΔT	Year	ΔT	Year	ΔT
1710	−0.2	1870	0.0	1940	+0.4
1730	−0.1	1880	−0.1	1950	+0.5
1750	0.0	1895	−0.1	1965	+0.6
1770	+0.1	1903	0.0	1971	+0.7
1800	+0.1	1912	+0.2	1977	+0.8
1840	0.0	1927	+0.4	1987	+1.0 ?

For epochs outside this time interval, an *approximate* value of ΔT (in minutes) can be calculated from

$$\Delta T = +0.41 + 1.2053\,T + 0.4992\,T^2 \tag{5.1}$$

where T is the time in centuries since 1900. We then have

$$\text{UT} = \text{ET} - \Delta T \qquad \text{or} \qquad \text{ET} = \text{UT} + \Delta T$$

Example 5.a :   Suppose that the position of Mercury should be calculated for February 6 at 6^h Universal Time of the year -555.

Here we have

$$T = -24.55, \quad \text{whence} \quad \Delta T = +272 \text{ minutes.}$$

Thus

$$ET = 6^h + 272 \text{ minutes} = 10^h 32^m$$

and the calculations should be performed for -555 February 6 at
$10^h 32^m$ ET.

Example 5.b :   According to the *Astronomical Ephemeris*, the maximum
phase of the lunar eclipse of 1977 April 4 took place at $4^h 19\overset{m}{.}0$
Ephemeris Time.

From Table 5.A, $\Delta T = +0.8$ minute in 1977. The corresponding UT
was thus

$$4^h 19\overset{m}{.}0 - 0\overset{m}{.}8 = 4^h 18\overset{m}{.}2.$$

6

Geocentric rectangular Coordinates
of an Observer

The figure represents a meridian cross-section of the Earth. C is the Earth's center, N its north pole, S its south pole, EF the equator, HK the horizontal plane of the observer O, and OP is perpendicular to HK. The direction OM, parallel to SN, makes with OH an angle ϕ which is the *geographical latitude* of O. The angle OPF too is equal to ϕ.

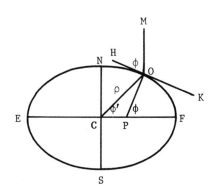

The radius vector OC, joining the observer to the center of the Earth, makes with the equator CF an angle ϕ' which is the *geocentric latitude* of O. We have $\phi = \phi'$ at the poles and at the equator ; for all other latitudes

$$|\phi'| < |\phi|$$

Let f be the Earth's flattening, and b/a the ratio NC/CF of the polar radius to the equatorial radius. With the value $f = 1/298.257$ now adopted by the International Astronomical Union, we have

$$\frac{b}{a} = 1 - f = 0.996\,647\,19$$

For a place at sea-level, we have

$$\tan \phi' = \frac{b^2}{a^2} \tan \phi$$

If H is the observer's height above sea-level in *meters*, the quantities $\rho \sin \phi'$ and $\rho \cos \phi'$, needed in the calculation of diurnal parallaxes, eclipses and occultations, may be calculated as follows :

$$\tan u = \frac{b}{a} \tan \phi$$

$$\left\{ \begin{array}{l} \rho \sin \phi' = \dfrac{b}{a} \sin u + \dfrac{H}{6\ 378\ 140} \sin \phi \\[2em] \rho \cos \phi' = \cos u + \dfrac{H}{6\ 378\ 140} \cos \phi \end{array} \right.$$

$\rho \sin \phi'$ is positive in the northern hemisphere, negative in the southern one, while $\rho \cos \phi'$ is always positive.

The quantity ρ denotes the observer's distance to the center of the Earth (*OC* in the Figure).

Exercise. – Calculate $\rho \sin \phi'$ and $\rho \cos \phi'$ for the Uccle Observatory, for which $\phi = +50°47'55''$ and $H = 105$ meters.

(Answer : $\rho \sin \phi' = +0.771\ 306$ and $\rho \cos \phi' = +0.633\ 333$).

7

Sidereal Time at Greenwich

The sidereal time at Greenwich at 0^h Universal Time of a given date can be obtained as follows.

Calculate the JD corresponding to that date at 0^h UT (see Chapter 3). Thus, this is a number ending on $.5$. Then find T by

$$T = \frac{JD - 2415\,020.0}{36525} \tag{7.1}$$

The sidereal time at Greenwich at 0^h UT, expressed in hours and decimals, is then

$$\theta_0 = 6.646\,0656 + 2400.051\,262\,T + 0.000\,025\,81\,T^2 \tag{7.2}$$

The result should be reduced to the interval $0 - 24$ hours, and then converted to hours, minutes and seconds if necessary.

To reduce θ_0 to the interval $0 - 24$ hours, it may be easier to divide the numerical values in formula (7.2) by 24 ; this gives

$$\theta_0 = 0.276\,919\,398 + 100.002\,1359\,T + 0.000\,001\,075\,T^2 \tag{7.3}$$

This gives the sidereal time in *revolutions*. Multiply the *fractional* part of the result by 24 in order to obtain θ_0 in hours.

It is important to note that the formulae (7.2) and (7.3) are valid only for those values of T which correspond to 0^h UT of a given date.

Example 7.a : Find the sidereal time at Greenwich on 1978 November 13 at 0^h Universal Time.

We find

$$JD = 2443\,825.5 \qquad T = +0.788\,651\,6085$$

and then, by formula (7.3),

$$\theta_0 = 79.143\,765\,40 \text{ revolution}$$
$$= 0.143\,765\,40 \text{ revolution}$$
$$= 3.450\,3696 \text{ hours}$$
$$= 3^h\,27^m\,01\overset{s}{.}331$$

The *A.E.* gives the same value.

To find the sidereal time *at Greenwich* for any instant UT of a given date, express that instant in hours and decimals, multiply by 1.002 737 908, and add the result to the sidereal time at 0^h UT.

Example 7.b : Find the sidereal time at Greenwich on 1978 November 13 at $4^h34^m00^s$ UT.

In the preceding example, we have found that the sidereal time at 0^h on that date is 3.450 3696 hours.

$$4^h34^m00^s = 4\overset{h}{.}566\,6667$$

$$4\overset{h}{.}566\,6667 \times 1.002\,737\,908 = 4\overset{h}{.}579\,1698$$

Hence, the required sidereal time is

$$\theta_0 = 3\overset{h}{.}450\,3696 + 4\overset{h}{.}579\,1698 = 8\overset{h}{.}029\,5394$$
$$= 8^h01^m46\overset{s}{.}342$$

The sidereal time obtained by formulae (7.2) or (7.3) is the *mean* sidereal time. The *apparent* sidereal time is obtained by adding the correction $\Delta\psi \cos\varepsilon$, where $\Delta\psi$ is the nutation in longitude (see Chapter 15), and ε the obliquity of the ecliptic. This correction for nutation is called *nutation in right ascension* (or *equation of the equinoxes* in the *A.E.*). The value of ε can here be taken to the nearest $10''$; if $\Delta\psi$ is expressed in seconds of a degree, the correction in seconds of time is

$$\frac{\Delta\psi \cos\varepsilon}{15}$$

Example 7.c : Find the apparent sidereal time at Greenwich on 1978 November 13 at $4^h34^m00^s$ UT.

From Example 7.b, the mean sidereal time at Greenwich for that instant is $8^h01^m46\overset{s}{.}342$, while $\Delta\psi = -3\overset{''}{.}378$ (see Example 15.a). Taking $\varepsilon = 23°26'30''$, the correction to the sidereal time is

$$\frac{-3.378 \times \cos\ 23°26'30''}{15} = -0.207 \text{ second}$$

and the required apparent sidereal time is

$$8^h 01^m 46^s.342 - 0^s.207 = 8^h 01^m 46^s.135$$

8

Transformation of Coordinates

We will use the following symbols :

α = right ascension. This quantity is generally expressed in hours, minutes and seconds, and thus should firstly be converted into degrees (and decimals) before to be used in a formula. Conversely, if α has been obtained by means of a formula and a calculating machine, it is expressed in degrees ; it may be converted to hours by division by 15, and then, if necessary, be converted into hours, minutes and seconds ;

δ = declination, positive (negative) if north (south) of the celestial equator ;

α_{1950} = right ascension referred to the standard equinox of 1950.0;

δ_{1950} = declination referred to the standard equinox of 1950.0 ;

λ = ecliptical (or celestial) longitude, measured from the vernal equinox along the ecliptic ;

β = ecliptical (or celestial) latitude, positive (negative) if north (south) of the ecliptic ;

l = galactic longitude ;

b = galactic latitude ;

h = altitude, positive (negative) if above (below) the horizon;

A = azimuth, measured westward from the *South*. It should be noted that several authors measure the azimuth from the North. We prefer to count it from the South, because the hour angles too are measured from the South. Thus, a celestial body which is exactly in the southern meridian has $A = H = 0°$;

ε = obliquity of the ecliptic ; this is the angle between the ecliptic and the celestial equator. The mean obliquity of the ecliptic is given by formula (18.4). If, however, the *apparent* right ascension and declination are used (that is, affected by the aberration and the nutation), the true obliquity $\varepsilon + \Delta\varepsilon$ should be used (see Chapter 15). If α and δ are referred to the standard equinox of 1950, then the value of ε for this epoch should be used, namely $\varepsilon_{1950} = 23°445\ 7889$. For the standard equinox of 2000.0, we have $\varepsilon_{2000} = 23°26'21\rlap{.}''448 = 23°439\ 2911$;

ϕ = the observer's latitude, positive (negative) if in the northern (southern) hemisphere ;

H = the local hour angle, measured westward from the South.

If θ is the local sidereal time, θ_0 the sidereal time at Greenwich, and L the observer's longitude (positive west, negative east from Greenwich), then the local hour angle can be calculated from

$$H = \theta - \alpha \qquad \text{or} \qquad H = \theta_0 - L - \alpha$$

If α is affected by the nutation, then the sidereal time too must be affected by it (see Chapter 7).

For the transformation from equatorial into ecliptical coordinates, the following formulae can be used :

$$\tan \lambda = \frac{\sin \alpha \cos \varepsilon + \tan \delta \sin \varepsilon}{\cos \alpha} \tag{8.1}$$

$$\sin \beta = \sin \delta \cos \varepsilon - \cos \delta \sin \varepsilon \sin \alpha \tag{8.2}$$

Transformation from ecliptical into equatorial coordinates :

$$\tan \alpha = \frac{\sin \lambda \cos \varepsilon - \tan \beta \sin \varepsilon}{\cos \lambda} \tag{8.3}$$

$$\sin \delta = \sin \beta \cos \varepsilon + \cos \beta \sin \varepsilon \sin \lambda \tag{8.4}$$

Calculation of the local horizontal coordinates :

$$\tan A = \frac{\sin H}{\cos H \sin \phi - \tan \delta \cos \phi} \tag{8.5}$$

$$\sin h = \sin \phi \sin \delta + \cos \phi \cos \delta \cos H \tag{8.6}$$

Transformation from equatorial coordinates, referred to the standard equinox of 1950.0, into galactic coordinates :

$$\tan x = \frac{\sin (192°25 - \alpha)}{\cos (192°25 - \alpha) \sin 27°4 - \tan \delta \cos 27°4} \qquad (8.7)$$

$$l = 303° - x$$

$$\sin b = \sin \delta \sin 27°4 + \cos \delta \cos 27°4 \cos (192°25 - \alpha) \qquad (8.8)$$

Transformation from galactic coordinates into equatorial coordinates referred to the standard equinox of 1950.0 :

$$\tan y = \frac{\sin (l - 123°)}{\cos (l - 123°) \sin 27°4 - \tan b \cos 27°4} \qquad (8.9)$$

$$\alpha = y + 12°25$$

$$\sin \delta = \sin b \sin 27°4 + \cos b \cos 27°4 \cos (l - 123°) \qquad (8.10)$$

Note the similitude of the formulae (8.1), (8.3), (8.5), (8.7) and (8.9). They can be calculated in one single subroutine. The same remark applies to the formulae (8.2), (8.4), (8.6), (8.8) and (8.10).

The formulae (8.1), (8.3), etc., give $\tan \lambda$, $\tan \alpha$, etc., and then λ, α, etc. by the function arctan. However, the exact quadrant where the angle is situated is then unknown. It is better *not* to calculate the tangent of the angle by making the division ; instead, apply the conversion from rectangular into polar coordinates on the numerator and the denominator of the fraction ; this will give the angle λ, α, etc. directly in the correct quadrant.

Example 8.a : Calculate the ecliptical coordinates of Pollux, whose equatorial coordinates are

$$\alpha_{1950} = 7^h 42^m 15^s 525, \qquad \delta_{1950} = +28°08'55''11.$$

Using the values $\alpha = 115°564\ 688$, $\delta = +28°148\ 642$, and $\varepsilon = 23°445\ 7889$, formulae (8.1) and (8.2) give

$$\tan \lambda = \frac{+1.040\ 5017}{-0.431\ 5299}, \qquad \text{whence} \quad \lambda = 112°525\ 38 ;$$

$$\beta = +6°68\ 058 .$$

Because α and δ are referred to the standard equinox of 1950.0, λ and β too are referred to that equinox.

Exercise. – Using the values of λ and β found in the preceding Example, find α and δ again by means of formulae (8.3) and (8.4).

Example 8.b : Find the azimuth and the altitude of Saturn on 1978 November 13 at $4^h34^m00^s$ UT at the Uccle Observatory (longitude $-0^h17^m25^s94$, latitude $+50°47'55''0$); the planet's apparent equatorial coordinates, interpolated from the *A.E.*, being

$$\alpha = 10^h57^m35^s681 \qquad \delta = +8°25'58''10$$

Since these are the *apparent* right ascension and declination, we need the *apparent* sidereal time. The latter has been calculated, for the given instant, in Example 7.c, namely $\theta_0 = 8^h01^m46^s135$. We thus have

$$H = \theta_0 - L - \alpha$$
$$= 8^h01^m46^s135 + 0^h17^m25^s94 - 10^h57^m35^s681$$
$$= -2^h38^m23^s606 = -2^h639\ 8906 = -39°598\ 358$$

Formulae (8.5) and (8.6) then give

$$\tan A = \frac{-0.637\ 4019}{+0.503\ 4048}, \quad \text{whence} \quad A = -51°6992$$

$$h = +36°5405$$

Exercise. – Find the galactic coordinates of Nova Serpentis 1978, whose equatorial coordinates are

$$\alpha_{1950} = 17^h48^m59^s74, \qquad \delta_{1950} = -14°43'08''2$$

(Answer : $l = 12°9593, \quad b = +6°0463$)

Rise or set of a body

The hour angle corresponding to the time of rise or set of a body is obtained by putting $h = 0$ in formula (8.6). This gives

$$\cos H = - \tan \phi \tan \delta$$

However, the instant so obtained refers to the geometric rise or set of the center of the celestial body.

By reason of the atmospheric refraction, the body is actually below the horizon at the instant of its apparent rise or set. The value of $0°34'$ is generally adopted for the effect of refraction at the horizon. For the Sun, the calculated times generally refer to the apparent rise or set of the upper limb of the disk ; hence, $0°16'$ should be added for the semidiameter. The hour angle H_o at the time of rise or set should thus be calculated from

$$\cos H_o = \frac{-0.00\ 989 - \sin \phi \sin \delta}{\cos \phi \cos \delta} \quad \text{for stars and planets;}$$

$$\cos H_o = \frac{-0.01\ 454 - \sin \phi \sin \delta}{\cos \phi \cos \delta} \quad \text{for the Sun.}$$

In the case of the Moon, the effect of the horizontal parallax also should be taken into account.

The value of $\cos H_o$ being given, there are two possible values for H_o :

$$-180° < H_o < 0° \qquad \text{for the rise,}$$

$$0° < H_o < +180° \quad \text{for the set.}$$

Pocket calculators generally give the value between $0°$ and $+180°$ by pressing the key for arcus cosinus (incorrectly labelled $\cos^{-1}$ on many machines). In that case, the sign of H_o should be changed in the case the time of rise is to be found. This can be performed by the use of a flag, which is set or cleared in the beginning of the program, and which later is interrogated.

The azimuth of a star at the time of its *geometric* rise or set is given by

$$\cos A_o = - \frac{\sin \delta}{\cos \phi}$$

where A_o should be taken between $180°$ and $360°$ (or between $-180°$ and $0°$) for the rise, and between $0°$ and $180°$ for the set.

Ecliptic and Horizon

If ε = obliquity of the ecliptic,
ϕ = latitude of the observer,
θ = local sidereal time,

then the longitudes of the two points of the ecliptic which are on the horizon are given by

$$\tan \lambda = \frac{-\cos \theta}{\sin \varepsilon \tan \phi + \cos \varepsilon \sin \theta} \qquad (8.11)$$

The angle I between the ecliptic and the horizon is given by

$$\cos I = \cos \varepsilon \sin \phi - \sin \varepsilon \cos \phi \sin \theta \qquad (8.12)$$

Example 8.c : For $\varepsilon = 23°44$, $\phi = +51°$, $\theta = 5^h00^m = 75°$, we find, from formula (8.11),

$\tan \lambda = -0.1879$, whence $\lambda = 169°21'$ and $\lambda = 349°21'$.

Formula (8.12) gives $I = 62°$.

Exercices

How does I vary in the course of a sidereal day ?

What happens with formula (8.11) when $\phi = 90° - \varepsilon$ and $\theta = 18^h$? Explain.

9

ANGULAR SEPARATION

The angular distance d between two celestial bodies, whose right ascensions and declinations are known, is given by the formula

$$\cos d = \sin \delta_1 \sin \delta_2 + \cos \delta_1 \cos \delta_2 \cos (\alpha_1 - \alpha_2) \tag{9.1}$$

where α_1 and δ_1 are the right ascension and the declination of one body, and α_2 and δ_2 those of the other one.

The same formula may be used when the ecliptical (celestial) longitudes λ and latitudes β of the two bodies are given, provided that α_1, α_2, δ_1 and δ_2 are replaced by λ_1, λ_2, β_1 and β_2.

Formula (9.1) may not be used when d is very near to $0°$ or $180°$ because in that case $|\cos d|$ is nearly equal to 1 and varies very slowly with d, so that d cannot be found accurately. For example :

$$\cos 0°01'00'' = 0.999\ 999\ 958$$
$$\cos 0°00'30'' = 0.999\ 999\ 989$$
$$\cos 0°00'15'' = 0.999\ 999\ 997$$
$$\cos 0°00'00'' = 1.000\ 000\ 000$$

If the angular separation is very small, say less than $0°10'$ or $0°05'$, this separation should be calculated from

$$d = \sqrt{(\Delta\alpha \cdot \cos \delta)^2 + (\Delta\delta)^2} \tag{9.2}$$

where $\Delta\alpha$ is the difference between the right ascensions, $\Delta\delta$ the difference between the declinations, while δ is the declination of any of the two bodies. It should be noted that $\Delta\alpha$ and $\Delta\delta$ should be expressed in the same angular units.

If $\Delta\alpha$ is expressed in hours (and decimals), $\Delta\delta$ in degrees (and decimals), then d expressed in seconds of a degree ($''$) is given by

$$d = 3600 \sqrt{(15 \Delta\alpha \cdot \cos \delta)^2 + (\Delta\delta)^2} \tag{9.3}$$

If $\Delta\alpha$ is expressed in seconds of time (s), and $\Delta\delta$ in seconds of a degree ($''$), then d expressed in $''$ is given by

$$d = \sqrt{(15\,\Delta\alpha.\cos\delta)^2 + (\Delta\delta)^2} \qquad (9.4)$$

Formulae (9.2), (9.3) and (9.4) may be used only when d is small.

Example 9.a : Calculate the angular distance between Arcturus (α Boo) and Spica (α Vir).

The 1950 coordinates of these stars are

α Boo : $\quad \alpha_1 = 14^h13^m22^s.8 = 213°.3450 \qquad \delta_1 = +19°26'31''$

α Vir : $\quad \alpha_2 = 13^h22^m33^s.3 = 200°.6388 \qquad \delta_2 = -10°54'03''$

Formula (9.1) gives $\cos d = +0.840\,342$,
whence $\quad d = 32°.8237 = 32°49'$

Exercise. – Calculate the angular distance between Aldebaran and Antares. (Answer : $169°58'$)

One or both bodies may be moving objects. For example : a planet and a star, or two planets. In that case, a program may be written where firstly the quantities δ_1, δ_2 and $(\alpha_1 - \alpha_2)$ are interpolated, after which d is calculated by means of one of the formulae (9.1) or (9.2). Hint : from the interpolated quantities, calculate $\cos d$ by means of formula (9.1). Then, if $\cos d < 0.999\,995$, find d ; if $\cos d > 0.999\,995$, use formula (9.2).

Exercise. – Using the following coordinates, calculate the instant and the value of the least distance between Mercury and Saturn.

1978 0 h ET	Mercury α_1	Mercury δ_1	Saturn α_2	Saturn δ_2
Sep 12	$10^h23^m17^s.65$	$+11°31'46''.3$	$10^h33^m01^s.23$	$+10°42'53''.5$
13	10 29 44.27	+11 02 05.9	10 33 29.64	+10 40 13.2
14	10 36 19.63	+10 29 51.7	10 33 57.97	+10 37 33.4
15	10 43 01.75	+ 9 55 16.7	10 34 26.22	+10 34 53.9
16	10 49 48.85	+ 9 18 34.7	10 34 54.39	+10 32 14.9

Answer : The least angular separation between the two planets is $0°03'44''$, on 1978 September 13 at $15^h06^m.5$ ET = 15^h06^m UT.

The same method can be used if one of the bodies is a star. The latter's coordinates are then constant. It is important to note that the α and δ of the star should be referred to the *same equinox* as that of the moving body.

If the moving body is a major planet, whose apparent right ascension and apparent declination referred to the equinox of the date are given (as in the *A.E.*, for instance), then for the star the apparent coordinates too must be used. If one takes the star's position from a catalogue, where they are referred to a standard equinox (for instance that of 1950.0), then the apparent α and δ are found by taking into account the proper motion of the star and the effects of precession, nutation and aberration.

If the α and δ of the moving body are referred to a standard equinox, then the α and δ of the star should be referred to this same standard equinox, the only corrections being those for the proper motion of the star.

10

CONJUNCTION BETWEEN TWO PLANETS

Given three or five ephemeris positions of two planets passing near each other, a program can be written which calculates the time of conjunction *in right ascension*, and the difference in declination between the two bodies at that time. The method consists of calculating the differences $\Delta\alpha$ of the corresponding right ascensions, and then calculating the time when $\Delta\alpha = 0$ by inverse interpolation, using formula (2.4) or (2.7). When that instant is found, direct interpolation of the differences $\Delta\delta$ of the declinations, by means of formula (2.3) or (2.5), yields the required difference in declination at the time of conjunction.

Example 10.a : Calculate the circumstances of the Mercury-Venus conjunction of 1979 November.

The following values, for 0^h ET of the date, are taken from the *Astronomical Ephemeris* :

1979		Mercury				Venus		
	α		δ		α		δ	
	h m s		$\circ$ $\prime$ $\prime\prime$		h m s		$\circ$ $\prime$ $\prime\prime$	
Nov. 7	16 11 38.61		−23 49 45.9		16 04 01.76		−21 07 49.3	
8	16 12 55.61		−23 46 54.0		16 09 14.55		−21 24 26.5	
9	16 13 40.37		−23 41 19.5		16 14 28.50		−21 40 27.5	
10	16 13 50.08		−23 32 50.8		16 19 43.58		−21 55 51.7	
11	16 13 22.16		−23 21 16.3		16 24 59.76		−22 10 38.4	

We firstly calculate the differences of the right ascensions (in hours and decimals) and of the declinations (in degrees and decimals) :

Nov.	$\Delta\alpha$	$\Delta\delta$
7	$= +0.126\,903$	$= -2.699\,06$
8	$+0.061\,406$	$-2.374\,31$
9	$-0.013\,369$	$-2.014\,44$
10	$-0.098\,194$	$-1.616\,42$
11	$-0.193\,778$	$-1.177\,19$

Applying formula (2.7) to the values of $\Delta\alpha$, we find that $\Delta\alpha$ is zero for the value $n = -0.16960$ of the interpolation interval. Hence, the conjunction in right ascension takes place on 1979 November 8.83040, that is on 1979 November 8 at 19^h55^m8 ET, or at 19^h55^m UT.

With the value of n just found, and applying formula (2.5) to the values of $\Delta\delta$, we find $\Delta\delta = -2°07808$ or $-2°05'$. Thus, at the time of conjunction in right ascension, Mercury is $2°05'$ south of Venus.

If the second body is a star, its coordinates may be considered as constant during the time interval considered. We then have

$$\alpha_1' = \alpha_2' = \alpha_3' = \alpha_4' = \alpha_5' \quad \text{and} \quad \delta_1' = \delta_2' = \delta_3' = \delta_4' = \delta_5'$$

The program can be written in such a manner that, if the second object is a star, its coordinates must be entered only once. To achieve this goal, use labels, flags and/or subroutines !

The important remark at the end of Chapter 9 does apply here too : *the coordinates of the star and those of the moving body must be referred to the same equinox.*

As an exercise, calculate the conjunction in right ascension between the minor planet 29 Amphitrite and the star λ Leonis in January 1980. The minor planet's right ascension and declination, referred to the standard equinox of 1950.0, are as follows (from an ephemeris calculated by David W. Dunham) :

0^h ET	α_{1950}	δ_{1950}
1980 January 7	$9^h34^m25^s279$	$+22°06'40''93$
12	9 31 10.656	+22 22 25.44
17	9 27 15.396	+22 39 04.68
22	9 22 45.672	+22 55 48.95
27	9 17 49.742	+23 11 46.00

The star's coordinates for the epoch and equinox of 1950.0 are $\alpha = 9^h28^m52^s248$ and $\delta = +23°11'22''21$, and the annual proper motion is -0^s0018 in right ascension and $-0''042$ in declination. Consequently, the star's position referred to the equinox of 1950.0 but for the epoch 1980.04 is

$$\alpha = 9^h28^m52^s194 , \qquad \delta = +23°11'20''95$$

Now, calculate the conjunction.

(Answer : Amphitrite passes $0°39'$ south of λ Leo on 1980 January 15 at $1\,h$).

11

BODIES IN STRAIGHT LINE

Let (α_1, δ_1), (α_2, δ_2), (α_3, δ_3) be the equatorial coordinates of three heavenly bodies. These three bodies are in "straight line" — that is, they lie on the same great circle of the celestial sphere — if

$$\tan \delta_1 \sin (\alpha_2 - \alpha_3) + \tan \delta_2 \sin (\alpha_3 - \alpha_1)$$
$$+ \tan \delta_3 \sin (\alpha_1 - \alpha_2) = 0 \qquad (11.1)$$

This formula is valid for ecliptical coordinates too, the right ascensions α being replaced by the longitudes λ, and the declinations δ by the latitudes β.

Do not forget that the right ascensions α are generally expressed in hours, minutes and seconds. They should firstly be converted into hours and decimals, and then into degrees by multiplication by 15.

If one or two of the bodies are stars, then once again the important remark at the end of Chapter 9 does apply : *the coordinates of the star(s) must be referred to the same equinox as that of the planet(s).*

Example 11.a : Find the time when Mars is seen in straight line with Pollux and Castor in 1979.

From an ephemeris of Mars and a star atlas, it is easily found that the planet is in straight line with the two stars about 1979 September 21. For this date, the apparent coordinates of the stars are :

Castor (α Gem) : $\quad \alpha_1 = 7^h33^m17^s\!.0 = 113°\!.3208$
$\qquad\qquad\qquad\qquad \delta_1 = +31°55'54'' = +31°\!.9317$

Pollux (β Gem) : $\quad \alpha_2 = 7^h44^m03^s\!.3 = 116°\!.0138$
$\qquad\qquad\qquad\qquad \delta_2 = +28°04'28'' = +28°\!.0744$

These values have been taken from the Soviet almanac *Astrono-micheskii Ezhegodnik* for 1979, pages 360 and 361, but they could have been calculated by means of the method described in Chapter 16. For our problem, these values of α_1, δ_1, α_2 and δ_2 may be considered as constants for several days.

The apparent coordinates of Mars (α_3, δ_3) are variable. Here are the values taken from the *Astronomical Ephemeris*

ET	α_3	δ_3
1979 Sep. 19.0	$7^h 54^m 33\overset{s}{.}8$ = 118°6408	+21°43'19″ = +21°7219
20.0	7 57 08.6 = 119.2858	+21 37 12 = +21.6200
21.0	7 59 42.7 = 119.9279	+21 30 57 = +21.5158
22.0	8 02 16.2 = 120.5675	+21 24 36 = +21.4100
23.0	8 04 49.0 = 121.2042	+21 18 08 = +21.3022

Using all these values, the first member of formula (11.1) takes the following values :

September 19.0	+0.002 1713
20.0	+0.001 2369
21.0	+0.000 3067
22.0	−0.000 6204
23.0	−0.001 5434

Using formula (2.7), we find that the value is zero for

1979 September 21.3304

= 1979 September 21 at 8^h ET (UT).

12

Smallest Circle
Containing Three Celestial Bodies

Let A, B, C be three celestial bodies situated not too far from each other on the celestial sphere, say closer than about 6°. We wish to calculate the angular diameter of the smallest circle containing these three bodies. Two cases can occur :

type I :   the smallest circle has as diameter the longest side of the triangle ABC ;

type II :   the smallest circle is the circle passing through the three points A, B, C.

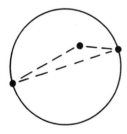

Type I

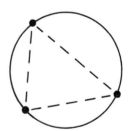

Type II

The diameter Δ of the smallest circle can be found as follows. Calculate the lengths of the three sides of the triangle ABC (in degrees) by means of formula (9.1). Formula (9.2) will rarely be required for the present problem.

Let a be the length of the *longest* side of the triangle, and b and c the lengths of the two other sides.

If $a > \sqrt{b^2 + c^2}$, then $\Delta = a$;

if $a < \sqrt{b^2 + c^2}$, then

$$\Delta = \frac{2\,abc}{\sqrt{(a + b + c)(a + b - c)(b + c - a)(a + c - b)}} \quad (12.1)$$

Example 12.a : Calculate the diameter of the smallest circle containing Mercury, Jupiter and Saturn on 1981 September 11 at $0\,h$ Ephemeris Time. The positions of these planets at that instant are :

Mercury	$\alpha = 12^h41^m08^s63$	$\delta = -5°37'54\rlap{.}''2$
Jupiter	12 52 05.21	-4 22 26.2
Saturn	12 39 28.11	-1 50 03.7

The three angular separations are found by means of formula (9.1):

Mercury - Jupiter	$3\rlap{.}°00152$
Mercury - Saturn	3.82028
Jupiter - Saturn	4.04599 $= a$

Because $4.04599 < \sqrt{(3.00152)^2 + (3.82028)^2} = 4.85836$, we calculate Δ by means of formula (12.1). The result is

$$\Delta = 4\rlap{.}°26364 = 4°16'$$

This is an example of type II.

Exercise. — Perform the same calculation for the planets Venus, Jupiter and Saturn on 1981 August 29 at 0^h ET, using the following positions :

Venus	$\alpha = 12^h46^m00^s82$	$\delta = -4°38'59\rlap{.}''7$
Jupiter	12 42 31.51	-3 20 36.0
Saturn	12 34 03.49	-1 14 18.2

Show that this case is of type I, and that $\Delta = 4°32'$.

A program can be written in which firstly the right ascensions and the declinations of the planets are interpolated, after which Δ is calculated. In that case, a test is necessary to compare a

with $\sqrt{b^2 + c^2}$. With such a program, it is possible to calculate (by trial) the minimum value of Δ of a grouping of three planets. Indeed, Δ varies with time, and the method described in this Chapter provides the value of Δ only for a given instant.

Such a program has been used by the author to calculate all the planetary "trios" occurring during the period 1960 - 2005. This list has been published in the French journal *L'Astronomie*, Volume 91, pages 487 - 493 (December 1977).

If one of the bodies is a star, once again the important remark at the end of Chapter 9 does apply : the coordinates of the star should be referred to the same equinox as those of the planets.

13

POSITION ANGLE OF THE MOON'S BRIGHT LIMB

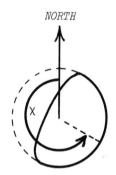

NORTH

The position angle of the Moon's bright limb is the position angle χ of the *midpoint* of the illuminated limb of the Moon, reckoned eastward from the North Point of the disk (see the Figure).

Let α and δ be the right ascension and the declination of the Sun, α' and δ' the right ascension and the declination of the Moon. Do not forget to express all these quantities in degrees and decimals !

Then the position angle χ of the Moon's bright limb is given by the formula

$$\tan \chi = \frac{\cos \delta \sin (\alpha - \alpha')}{\cos \delta' \sin \delta - \sin \delta' \cos \delta \cos (\alpha - \alpha')}$$

The angle χ is in the vicinity of 270° near First Quarter, near 90° after Full Moon. However, χ is found immediately in the correct quadrant by applying the conversion from rectangular to polar coordinates to the numerator and the denominator of the fraction in the preceding formula.

Example 13.a : Find the position angle of the Moon's bright limb on 1979 February 2 at 21^h ET.

The Moon's equatorial coordinates for that instant are given on page 78 of the *A.E.* for 1979 :

α' = 1^{h}54^{m}18^{s}175 = 1^{h}90505 = 28°5757
δ' = +8°01'47"59 = +8°0299

The coordinates of the Sun are found, by interpolation, from the values given on page 21 of the same publication :

$$\alpha = 21^h\!.05953 = 315°\!.8930$$
$$\delta = -16°\!.7915$$

We then find

$$\tan \chi = \frac{-0.91397}{-0.32586}$$

from which $\chi = -109°\!.62 = 250°\!.38$.

When the position angle of (the midpoint of) the Moon's bright limb is known, it's easy to see what part of the lunar limb is illuminated by the Sun. It's simply the arc from $\chi - 90°$ to $\chi + 90°$. In Example 13.a, we have found $\chi = 250°$. Hence, the illuminated limb of the Moon is that from position angle 160° to position angle 340°, in that case.

Of course, if $\chi = 283°$, for instance, the illuminated limb of the Moon is the arc from position angle 193° through 283° to 360° and further to 13°.

14

Precession

In this Chapter, we consider the problem of converting the right ascension α and the declination δ of a star, given for an epoch and an equinox, to the corresponding values for another epoch and equinox. Only the *mean* places of a star and the effect of the precession alone are considered here. The problem of finding the apparent place of a star will be considered in Chapter 16.

If no great accuracy is required, and if the two epochs are not widely separated, the following formulae may be used for the annual precessions in right ascension and declination :

$$\Delta\alpha = m + n \sin \alpha \tan \delta \qquad \Delta\delta = n \cos \alpha \qquad (14.1)$$

where m and n are two quantities which vary slowly with time. They are given by

$$m = 3\overset{s}{.}07234 + 0\overset{s}{.}00186\,T\,,$$
$$n = 20\overset{''}{.}0468 - 0\overset{''}{.}0085\,T\,,$$

T being the time measured in centuries from 1900.0. Here are the values of m and n for some epochs :

Epoch	m	n	n
			$''$
1700.0	$3\overset{s}{.}069$	$1\overset{s}{.}338$	20.06
1800.0	3.070	1.337	20.06
1900.0	3.072	1.336	20.05
2000.0	3.074	1.336	20.04
2100.0	3.076	1.335	20.03
2200.0	3.078	1.335	20.02

For the calculation of $\Delta\alpha$, the value of n expressed in seconds of time (s) must be used. Remember that 1^s corresponds to $15''$.

The effect of the proper motion should be added to the values given by the formulae (14.1).

Example 14.a :   The coordinates of Regulus for the epoch and equinox of 1950.0 are

$$\alpha_0 = 10^h05^m42^s7 \qquad\qquad \delta_0 = +12°12'45''$$

and the annual proper motions are

        -0^s0171 in right ascension,
        $+0''004$ in declination.

Reduce these coordinates to the epoch and the equinox of 1978.0.

Here we have

$$\alpha = 151°428 \qquad\qquad \delta = +12°213$$
$$m = 3^s073 \qquad\qquad n = 1^s336 = 20''04$$

From the formulae (14.1) we deduce

$$\Delta\alpha = +3^s211, \qquad\qquad \Delta\delta = -17''60$$

to which we must add the annual proper motion, giving an annual variation of $+3^s194$ in right ascension, and of $-17''60$ in declination.

    Variations during 28 years (from 1950.0 to 1978.0) :

       in α : $+3^s194 \times 28$ $=$ $+89^s4$ $=$ $+1^m29^s4$

       in δ : $-17''60 \times 28$ $=$ $-493''$ $=$ $-8'13''$

Required right ascension : $\alpha = \alpha_0 + 1^m29^s4 = 10^h07^m12^s1$

Required declination : $\delta = \delta_0 - 8'13'' = +12°04'32''$

The *A.E.* for 1978, page 336, gives $10^h07^m12^s1$ and $+12°04'31''$.

Rigorous method

Newcomb gives the following numerical expressions for the quantities ζ, z and θ which are needed for the accurate reduction of positions from one equinox to another :

$$\text{Initial epoch :} \quad t_0 = 1900.0 + \tau_0$$

$$\text{Final epoch :} \quad t = 1900.0 + \tau_0 + \tau$$

$$\left.\begin{aligned}
\zeta &= (2304\rlap{.}''250 + 1\rlap{.}''396\,\tau_0)\,\tau + 0\rlap{.}''302\,\tau^2 + 0\rlap{.}''018\,\tau^3 \\
z &= \zeta + 0\rlap{.}''791\,\tau^2 + 0\rlap{.}''001\,\tau^3 \\
\theta &= (2004\rlap{.}''682 - 0\rlap{.}''853\,\tau_0)\,\tau - 0\rlap{.}''426\,\tau^2 - 0\rlap{.}''042\,\tau^3
\end{aligned}\right\} \qquad (14.2)$$

where τ_0 and τ are measured in *tropical* centuries of 36524.2199 ephemeris days. The fundamental epoch 1900.0 corresponds to JD 2415 020.313. The length of the tropical year is slightly variable with time by about −0.53 second per century, but this very small decrease may be neglected for our purpose.

In other words, if $(JD)_0$ and (JD) are the Julian Days corresponding to the initial and the final epoch, respectively, we have

$$\tau_0 = \frac{(JD)_0 - 2415\,020.313}{36524.2199} \qquad\qquad \tau = \frac{(JD) - (JD)_0}{36524.2199}$$

For $t_0 = 1950.0 = $ JD 2433 282.423, we have $\tau_0 = +0.5$ and the expressions (14.2) become

$$\left.\begin{aligned}
\zeta &= 2304\rlap{.}''948\,\tau + 0\rlap{.}''302\,\tau^2 + 0\rlap{.}''018\,\tau^3 \\
z &= 2304\rlap{.}''948\,\tau + 1\rlap{.}''093\,\tau^2 + 0\rlap{.}''019\,\tau^3 \\
\theta &= 2004\rlap{.}''255\,\tau - 0\rlap{.}''426\,\tau^2 - 0\rlap{.}''042\,\tau^3
\end{aligned}\right\} \qquad (14.3)$$

Then, the rigorous formulae for the reduction of the given equatorial coordinates α_0 and δ_0 of the epoch t_0 to the coordinates α and δ of the epoch t are :

$$A = \cos\delta_0 \sin(\alpha_0 + \zeta)$$

$$B = \cos\theta \cos\delta_0 \cos(\alpha_0 + \zeta) - \sin\theta \sin\delta_0$$

$$C = \sin\theta \cos\delta_0 \cos(\alpha_0 + \zeta) + \cos\theta \sin\delta_0$$

$$\tan (\alpha - z) = \frac{A}{B} \qquad \qquad \sin \delta = C$$

Apply the rectangular/polar coordinates transformation to the quantities A and B. This will give $(\alpha - z)$ directly in the correct quadrant, and also give $\cos \delta = \sqrt{A^2 + B^2}$ which may be used instead of $\sin \delta$ if the star is very close to the pole.

Before making the reduction from α_o, δ_o to α, δ, calculate the effect of the star's proper motion.

Example 14.b : The star θ Persei has the following mean coordinates for the epoch and equinox of 1950.0 :

$$\alpha_o = 2^h 40^m 46\overset{s}{.}276 \qquad \qquad \delta_o = +49°01'06\overset{''}{.}45$$

and its annual proper motions referred to that same equinox are

$$+0\overset{s}{.}0342 \quad \text{in right ascension,}$$
$$-0\overset{''}{.}083 \quad \text{in declination.}$$

Reduce the coordinates to the epoch and mean equinox of 1978 November 13.19 UT.

The initial epoch is 1950.0 or JD 2433 282.423, and the final one is JD 2443 825.69. Hence, τ = +0.288 665 tropical centuries, or 28.8665 years.

We firstly calculate the effect of the proper motion. The variations over 28.8665 years are

$$+0\overset{s}{.}0342 \times 28.8665 = +0\overset{s}{.}987 \quad \text{in right ascension,}$$
$$-0\overset{''}{.}083 \times 28.8665 = -2\overset{''}{.}40 \quad \text{in declination.}$$

Thus the star's coordinates, for the mean equinox of 1950.0, but for the epoch 1978 November 13.19, are

$$\alpha_o = 2^h 40^m 46\overset{s}{.}276 + 0\overset{s}{.}987 = 2^h 40^m 47\overset{s}{.}263 = +40\overset{\circ}{.}196\ 929$$

$$\delta_o = +49°01'06\overset{''}{.}45 - 2\overset{''}{.}40 = +49°01'04\overset{''}{.}05 = +49\overset{\circ}{.}017\ 792$$

Since the initial equinox is that of 1950.0, we can use the formulae (14.3). With the value τ = +0.288 665, we obtain

$$\zeta = +665\overset{''}{.}383 = +0\overset{\circ}{.}184\ 829$$
$$z = +665\overset{''}{.}449 = +0\overset{\circ}{.}184\ 847$$
$$\theta = +578\overset{''}{.}522 = +0\overset{\circ}{.}160\ 701$$

$$A = +0.424\ 893\ 97$$
$$B = +0.497\ 451\ 58$$
$$C = +0.756\ 311\ 48$$

$$\alpha - z = +40\overset{\circ}{.}502\ 010$$
$$\alpha = +40\overset{\circ}{.}686\ 857 = 2^h42^m44\overset{s}{.}846$$
$$\delta = +49\overset{\circ}{.}140\ 096 = +49°08'24''35$$

Exercise. − For the same star as in Example 14.b, calculate the equatorial coordinates for the epoch and mean equinox of 1981.0.

Answer : Here, $\tau = +0.31$, and one finds
$$\alpha = 2^h42^m53\overset{s}{.}626, \qquad \delta = +49°08'56''58.$$

Exercise. − The equatorial coordinates of α Ursae Minoris, for the epoch and mean equinox of 1950.0, are
$$\alpha = 1^h48^m48\overset{s}{.}786, \qquad \delta = +89°01'43''74$$

and the star's annual proper motions for the same equinox are

$$+0\overset{s}{.}1811 \quad \text{in right ascension,}$$
$$-0''004 \quad \text{in declination.}$$

Find the coordinates of the star for the epochs and mean equinoxes of 1800.0, 1980.0 and 2100.0.

Answer :

	α	δ
1800.0	$0^h52^m25\overset{s}{.}31$	$+88°14'24''52$
1980.0	2 11 47.60	+89 10 24.41
2100.0	5 53 33.88	+89 32 21.81

It should be noted that the formulae (14.2) are valid only for a limited period of time. If we use them for the year 32 600, for instance, we find for that epoch that α UMi will be at declination −87°, a completely wrong result !

15

NUTATION

The nutation in longitude ($\Delta\psi$) and the nutation in obliquity ($\Delta\epsilon$) are needed for the calculation of the apparent place of a star and for that of the apparent sidereal time. For a given instant, $\Delta\psi$ and $\Delta\epsilon$ can be calculated as follows.

Find the time T, measured in Julian centuries from 1900 January 0.5, by means of the formula

$$T = \frac{\text{JD} - 2415\,020.0}{36525} \tag{15.1}$$

where JD is the Julian Day (see Chapter 3). Then calculate the angles L, L', M, M' and Ω by means of the following formulae, in which the various constants are expressed in degrees and decimals. If T is small or when no high accuracy is required, the terms in T^2 may be neglected.

Sun's mean longitude :

$L = 279.6967 + 36000.7689\,T + 0.000\,303\,T^2$

Moon's mean longitude :

$L' = 270.4342 + 481\,267.8831\,T - 0.001\,133\,T^2$

Sun's mean anomaly :

$M = 358.4758 + 35999.0498\,T - 0.000\,150\,T^2$

Moon's mean anomaly :

$M' = 296.1046 + 477\,198.8491\,T + 0.009\,192\,T^2$

Longitude of Moon's ascending node :

$\Omega = 259.1833 - 1934.1420\,T + 0.002\,078\,T^2$

We then have, neglecting smaller quantities, and the coefficients being expressed in seconds of a degree ($''$) :

$$
\begin{aligned}
\Delta\psi = \ & - \ (17.2327 + 0.01737\,T) \sin \Omega \\
& - \ (1.2729 + 0.00013\,T) \sin 2L \\
& + \ 0.2088 \sin 2\Omega \\
& - \ 0.2037 \sin 2L' \\
& + \ (0.1261 - 0.00031\,T) \sin M \\
& + \ 0.0675 \sin M' \\
& - \ (0.0497 - 0.00012\,T) \sin (2L + M) \\
& - \ 0.0342 \sin (2L' - \Omega) \\
& - \ 0.0261 \sin (2L' + M') \\
& + \ 0.0214 \sin (2L - M) \\
& - \ 0.0149 \sin (2L - 2L' + M') \\
& + \ 0.0124 \sin (2L - \Omega) \\
& + \ 0.0114 \sin (2L' - M')
\end{aligned}
$$

$$
\begin{aligned}
\Delta\varepsilon = \ & + \ (9.2100 + 0.00091\,T) \cos \Omega \\
& + \ (0.5522 - 0.00029\,T) \cos 2L \\
& - \ 0.0904 \cos 2\Omega \\
& + \ 0.0884 \cos 2L' \\
& + \ 0.0216 \cos (2L + M) \\
& + \ 0.0183 \cos (2L' - \Omega) \\
& + \ 0.0113 \cos (2L' + M') \\
& - \ 0.0093 \cos (2L - M) \\
& - \ 0.0066 \cos (2L - \Omega)
\end{aligned}
$$

If no high accuracy is required, the smaller terms and the terms in T may be neglected. In the expressions for $\Delta\psi$ and $\Delta\varepsilon$, the first term has a period of 6798 days (18.61 years), and the second term a period of 182.62 days.

Example 15.a : Calculate $\Delta\psi$ and $\Delta\varepsilon$ for 1978 November 13 at 4^h35^m Ephemeris Time, that is for 4^h34^m Universal Time.

We find successively :

JD $= 2443\ 825.69$	$M' = 376\ 642°2324 = 82°2324$
$T = +0.788\ 656\ 810$	$\Omega = -1266°1897 = +173°8103$
$L = 28\ 671°9485 = 231°9485$	
$L' = 379\ 825°6269 = 25°6269$	$\Delta\psi = -3''378$
$M = 28\ 749°3715 = 309°3715$	$\Delta\varepsilon = -9''321$

According to the *A.E.* the correct values are $-3''383$ and $-9''321$, respectively.

16

Apparent Place of a Star

The *mean place* of a star at any time is its apparent position on the celestial sphere, as it would be seen by an observer at rest on the Sun, and referred to the ecliptic and mean equinox of the date (or to the mean equator and mean equinox of the date).

The *apparent* place of a star at any time is its position on the celestial sphere as it is actually seen from the center of the moving Earth, and referred to the instantaneous equator, ecliptic and equinox. It should be noted that :

- the *mean equinox* is the intersection of the ecliptic of date with the mean equator ;

- the *true equinox* is the intersection of the ecliptic of date with the true equator (that is, the equator affected by the nutation) ;

- there is no "mean" ecliptic, because the ecliptic has a regular motion.

The problem of the reduction of the place of a star from the mean place at one time (for instance of a standard epoch and equinox) to the apparent place of another time, involves the following corrections :

(A) The *proper motion* of the star between the two epochs. We may assume that by its proper motion each star moves on a great circle with an invariable angular speed. Except when the proper motion is an important fraction of the polar distance of the star, not only the proper motion itself, but also its components in right ascension and declination *with respect to a fixed equinox* may be considered as constants during several centuries. Therefore, we start by finding the effect of the proper motion when the axes of reference remain fixed, as in Example 14.b ;

(B) The effect of *precession*. This has been explained in Chapter 14;

(C) The effect of *nutation* (see below) ;

(D) The effect of *annual aberration* (see below) ;

(E) The effect of the *annual parallax*. This correction never exceeds $0''.8$, and may be neglected in most cases.

The changes in right ascension and in declination due to the *nutation* are

$\Delta\alpha_1 = (\cos \varepsilon + \sin \varepsilon \sin \alpha \tan \delta) \Delta\psi - (\cos\alpha \ \tan\delta) \Delta\varepsilon$

$\Delta\delta_1 = (\sin \varepsilon \cos \alpha) \Delta\psi + (\sin \alpha) \Delta\varepsilon$

The quantities $\Delta\psi$ and $\Delta\varepsilon$ may be calculated by means of the method described in Chapter 15, or be taken from the *A.E.*, while ε is the obliquity of the ecliptic, given by formula (18.4).

If Θ is the true longitude of the Sun, which can be calculated by means of the method described in Chapter 18, the changes in right ascension and in declination of a star due to the *annual aberration* are

$$\Delta\alpha_2 = -20''.49 \ \frac{\cos \alpha \cos \Theta \cos \varepsilon + \sin \alpha \sin \Theta}{\cos \delta}$$

$$\Delta\delta_2 = -20''.49 \left[\cos \Theta \cos\varepsilon \ (\tan \varepsilon \cos\delta - \sin\alpha \sin\delta) \right.$$
$$\left. + \cos \alpha \sin \delta \sin \Theta \right]$$

where, as above, α and δ are the star's right ascension and declination.

The total corrections to α and δ are therefore $\Delta\alpha_1 + \Delta\alpha_2$ and $\Delta\delta_1 + \Delta\delta_2$, respectively. Calculated from the above formulae, both are expressed in seconds of a degree. Divide the correction to α by 15 in order to obtain it in seconds of time.

Example 16.a : Calculate the apparent place of θ Persei for 1978 November 13.19 UT.

The mean position of this star for that instant, including the effect of proper motion, was found in Example 14.b, namely

$\alpha = 2^h42^m44^s.846 = 40°.687 \qquad \delta = +49°08'24''.35 = +49°.140$

The nutations in longitude and in obliquity, for the same instant, were found in Example 15.a :

$$\Delta\psi = -3''378 \qquad \Delta\varepsilon = -9''321$$

The Sun's true longitude, calculated by the method of Chapter 18, is $\Theta = 230°45$, while $\varepsilon = 23°44$. (For both values, an accuracy of 0.01 degree is sufficient in this case).

Putting the values of α, δ, ε, Θ, $\Delta\psi$ and $\Delta\varepsilon$ in the above-given formulae, one finds

$$\Delta\alpha_1 = +4''059 \qquad \Delta\delta_1 = -7''096$$
$$\Delta\alpha_2 = +29''619 \qquad \Delta\delta_2 = +6''554$$

and the total corrections in right ascension and declination are

$$\Delta\alpha = +4''059 + 29''619 = +33''678 = +2^s245$$
$$\Delta\delta = -7''096 + 6''554 = -0''54$$

Hence, the required apparent coordinates are

$$\alpha = 2^h42^m44^s846 + 2^s245 = 2^h42^m47^s09$$
$$\delta = +49°08'24''35 - 0''54 = +49°08'23''8$$

The values interpolated from the data on page 321 of the *Astronomicheskii Ezhegodnik 1978* are

$$2^h42^m47^s100 \qquad \text{and} \qquad +49°08'23''86$$

17

REDUCTION OF ECLIPTICAL ELEMENTS
FROM ONE EQUINOX TO ANOTHER ONE

For some problems, it may be necessary to reduce orbital elements
of a planet, a minor planet or a comet from one equinox to another
one. Of course, the semimajor axis a and the eccentricity e do not
change when the orbit is referred to another equinox, and thus
only the three elements

i = inclination,
ω = argument of perihelion,
Ω = longitude of ascending node

should be taken into consideration here. Let i_0, ω_0, Ω_0 be the
known values of these elements at the initial epoch τ_0, and i, ω,
Ω their (unknown) values at the final epoch τ. If τ_0 and τ are expressed in *thousands* of tropical years since 1900.0, and if

$$t = \tau - \tau_0$$

calculate the following values :

$$\eta = (471\rlap{.}{''}07 - 6\rlap{.}{''}75\,\tau_0 + 0\rlap{.}{''}57\,\tau_0^2)t + (-3\rlap{.}{''}37 + 0\rlap{.}{''}57\,\tau_0)t^2 + 0\rlap{.}{''}05\,t^3$$

$$\theta_0 = 173\rlap{.}{°}950833 + 32869''\tau_0 + 56''\tau_0^2 - (8694'' + 55''\tau_0)t + 3''t^2$$

$$\theta = \theta_0 + (50256\rlap{.}{''}41 + 222\rlap{.}{''}29\,\tau_0 + 0\rlap{.}{''}26\,\tau_0^2)t + (111\rlap{.}{''}15 + 0\rlap{.}{''}26\,\tau_0)t^2 + 0\rlap{.}{''}1\,t^3$$

In the Figure, E_0 and γ_0 are the ecliptic and the vernal equinox
at epoch τ_0, and E and γ the ecliptic and equinox at epoch τ. The
angle between the two ecliptics is η. The orbit's perihelion is
denoted by Π.

Then the quantities i and $\Omega - \theta$, and thus Ω, can be calculated
from

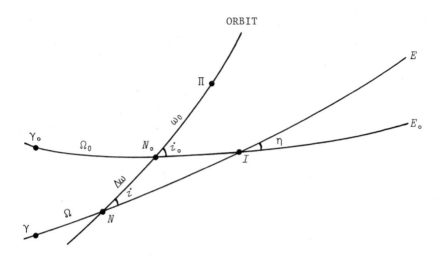

$$\cos i = \cos i_o \cos \eta + \sin i_o \sin \eta \cos (\Omega_o - \theta_o) \qquad (17.1)$$

$$\sin i \sin (\Omega - \theta) = \sin i_o \sin (\Omega_o - \theta_o) \qquad (17.2)$$
$$\sin i \cos (\Omega - \theta) = -\sin \eta \cos i_o + \cos \eta \sin i_o \cos (\Omega_o - \theta_o)$$

Formula (17.1) should not be used if the inclination is small. Then $\omega = \omega_o + \Delta\omega$, where $\Delta\omega$ is found from

$$\sin i \sin \Delta\omega = -\sin \eta \sin (\Omega_o - \theta_o)$$
$$\sin i \cos \Delta\omega = \sin i_o \cos \eta - \cos i_o \sin \eta \cos (\Omega_o - \theta_o) \qquad (17.3)$$

If $i_o = 0$, Ω is not determined. Then $i = \eta$ and $\Omega = \theta + 180°$.

Example 17.a : In their *Catalogue Général des Orbites de Comètes de l'an −466 à 1952*, F. Baldet and G. De Obaldia give the following orbital elements for comet Klinkenberg (1744), referred to the mean equinox of 1744.0 :

$$i_o = 47°1220$$
$$\omega_o = 151.4486$$
$$\Omega_o = 45.7481$$

Reduce these elements to the standard equinox of 1950.0.

We find successively :

$$\tau_o = \frac{1744 - 1900}{1000} = -0.156 \qquad \tau = \frac{1950 - 1900}{1000} = +0.050$$

$$t = +0.206$$

$$\eta = +97''114 = +0°026\,9761$$

$$\theta_o = 173°950\,833 - 6915''270 = 172°029\,925$$

$$\theta = 172°029\,925 + 10350''394 = 174°905\,035$$

Then formulae (17.2) give

$$\sin i \sin (\Omega - \theta) = -0.5907\,2524$$
$$\sin i \cos (\Omega - \theta) = -0.4339\,6271$$

whence, using the key for transformation from rectangular to polar coordinates,

$$\sin i = +0.7329\,9382 \qquad \text{whence} \quad i = 47°1380$$
$$\Omega - \theta = -126°3020 \qquad \text{whence} \quad \Omega = 48°6030$$

Formulae (17.3) give

$$\sin i \sin \Delta\omega = +0.0003\,7954$$
$$\sin i \cos \Delta\omega = +0.7329\,9372$$

whence $\Delta\omega = +0°0297$, and $\omega = 151°4783$.

In his *Catalogue of Cometary Orbits* (1975), B.G. Marsden gives the values $i = 47°1378$, $\omega = 151°4783$, $\Omega = 48°6030$.

18

Solar Coordinates

Let JD be the Julian (Ephemeris) Date, which can be calculated by means of the method described in Chapter 3. Then the time T, measured in Julian centuries of 36525 ephemeris days from the epoch 1900 January 0.5 ET, is given by

$$T = \frac{JD - 2415\,020.0}{36525} \qquad (18.1)$$

This quantity should be calculated with a sufficient number of decimals. For instance, five decimals are not sufficient (unless the Sun's longitude is required with an accuracy not better than one degree) : remember that T is expressed in centuries, so that an error of 0.00001 in T corresponds to an error of 0.37 day in the time.

Then the geometric mean longitude of the Sun, referred to the mean equinox of the date, is given by

$$L = 279°69668 + 36000°76892\,T + 0°000\,3025\,T^2$$

The Sun's mean anomaly is

$$M = 358°47583 + 35999°04975\,T - 0°000\,150\,T^2 - 0°000\,0033\,T^3$$

The eccentricity of the Earth's orbit is

$$e = 0.016\,751\,04 - 0.000\,0418\,T - 0.000\,000\,126\,T^2$$

To find the Sun's true longitude and true anomaly, two different methods can be used.

FIRST METHOD : With the values of M and e, solve Kepler's equation to find the eccentric anomaly E, using one of the methods described in Chapter 22. Then calculate the true anomaly v by means of Formula (25.1).

The Sun's true longitude $\odot$ is then

$$\odot = L + v - M$$

SECOND METHOD : Calculate the Sun's equation of the center C as follows :

$$C = + (1°919\ 460 - 0°004\ 789\ T - 0°000\ 014\ T^2)\ \sin M$$
$$+ (0°020\ 094 - 0°000\ 100\ T)\ \sin 2M$$
$$+ 0°000\ 293\ \sin 3M$$

Then the Sun's true longitude is

$$\odot = L + C$$

and its true anomaly is $v = M + C$.

The Sun's radius vector, expressed in astronomical units, can then be obtained by means of one of the following expressions :

$$R = 1.000\ 0002\ (1 - e \cos E)$$

$$R = \frac{1.000\ 0002\ (1 - e^2)}{1 + e \cos v} \qquad (18.2)$$

In the second formula, the numerator is a quantity which varies slowly with time. It is equal to

0.999 7182	in the year	1800
0.999 7196		1900
0.999 7210		2000
0.999 7224		2100

The Sun's longitude $\odot$, obtained by the method described above, is the true *geometric* longitude referred to the *mean* equinox of the date. This longitude is the quantity required for instance in the calculation of geocentric planetary positions.

If the *apparent* longitude of the Sun, referred to the *true* equinox of the date, is required, it is necessary to correct $\odot$ for the nutation and the aberration. Unless high accuracy is required, this can be performed as follows.

$$\Omega = 259°18 - 1934°142\ T$$

$$\odot_{app} = \odot - 0°00569 - 0°00479\ \sin \Omega$$

In some instances, for example in meteor work, it is necessary to have the Sun's longitude referred to the standard equinox of 1950.0. For the 20th century, this can be performed with a sufficient accuracy as follows :

$$\Theta_{1950} = \Theta - 0^\circ\!\!.01396 \, (year - 1950)$$

The Sun's latitude is ever less than $1''\!\!.2$, and may thus be put equal to zero unless high accuracy is required. In that case, the Sun's right ascension α and declination δ can be calculated from

$$\tan \alpha = \frac{\cos \varepsilon \sin \Theta}{\cos \Theta} \qquad (18.3)$$

$$\sin \delta = \sin \varepsilon \sin \Theta$$

where ε, the obliquity of the ecliptic, is given by

$$\begin{aligned}
\varepsilon = 23^\circ\!\!.452\,294 &- 0^\circ\!\!.013\,0125\,T \\
&- 0^\circ\!\!.000\,001\,64\,T^2 \qquad (18.4) \\
&+ 0^\circ\!\!.000\,000\,503\,T^3
\end{aligned}$$

If the *apparent* position of the Sun is required, ε should be corrected by

$$+ 0^\circ\!\!.00\,256 \cos \Omega \qquad (18.5)$$

Formula (18.3) may of course be transformed to

$$\tan \alpha = \cos \varepsilon \tan \Theta$$

and then it must be remembered that α must be in the same quadrant as Θ. However, for programmable pocket calculators it is better to leave formula (18.3) unchanged, and to apply the rectangular/polar coordinate conversion to the quantities $\cos \varepsilon \sin \Theta$ and $\cos \Theta$.

The value found for α will be expressed in degrees. Divide the result by 15 in order to express it into hours.

Higher accuracy

A somewhat better accuracy can be obtained as follows. Calculate the angles A, B, C, D, E, H by means of the following expressions, where all numerical values are in degrees and decimals.

$$A = 153.23 + 22518.7541\ T$$
$$B = 216.57 + 45037.5082\ T$$
$$C = 312.69 + 32964.3577\ T$$
$$D = 350.74 + 445\,267.1142\ T - 0.00144\ T^2$$
$$E = 231.19 + 20.20\ T$$
$$H = 353.40 + 65928.7155\ T$$

Then add the following corrections to the Sun's longitude :

$$+ 0°\!.00134 \cos A$$
$$+ 0°\!.00154 \cos B$$
$$+ 0°\!.00200 \cos C$$
$$+ 0°\!.00179 \sin D$$
$$+ 0°\!.00178 \sin E$$

and the following corrections to the radius vector :

$$+ 0.000\,005\,43 \sin A$$
$$+ 0.000\,015\,75 \sin B$$
$$+ 0.000\,016\,27 \sin C$$
$$+ 0.000\,030\,76 \cos D$$
$$+ 0.000\,009\,27 \sin H$$

The terms involving A and B are due to the action of Venus, the terms with arguments C and H are due to Jupiter, the terms with D are due to the Moon, while the term involving E is an inequality of long period.

Example 18.a :   Calculate the Sun's position on 1978 November 12 at 0^h ET = JD 2443 824.5.

We find successively :

$$T = +0.788\ 624\ 230$$
$$L = 28670°77554 = 230°77554$$
$$M = 28748°19863 = 308°19863$$
$$e = 0.016\ 718\ 00$$

With these values for M and e, the solution of Kepler's equation (see Chapter 22) is $E = 307°43807$. Then we find, using formula (25.1), $v = 306°67358$.

The Sun's true longitude is then

$$\Theta = L + v - M = 229°25049 = 229°15'02''$$

Using the second method, we find that the equation of the center is

$$C = 1°915\ 6746 \sin M + 0°020\ 0151 \sin 2M$$
$$+ 0°000\ 293 \sin 3M$$
$$= -1°52505$$

whence

$$\Theta = L + C = 229°25049,\quad \text{the same result as above.}$$

Then each of the formulae (18.2) gives $R = 0.98984$.

The correct values, according to the *A.E.*, are

$$\Theta = 229°15'05''85 \quad\text{and}\quad R = 0.989\ 8375.$$

If the apparent longitude of the Sun is required, we have $\Omega = -1266°13 = +173°87$, whence

$$\Theta_{app} = 229°25049 - 0°00569 - 0°00479 \sin 173°87$$
$$= 229°24429 = 229°14'39''.$$

According to the *A.E.*, the correct value is $229°14'41''86$.

Using formulae (18.4) and (18.5), we find $\varepsilon = 23°43949$, from which we deduce, using $\Theta_{app} = 229°24429$,

$$\alpha = -133°20853 = +226°79147 = 15^h119431 = 15^h07^m10^s0$$
$$\delta = -17°53682 = -17°32'13''$$

The *A.E.* gives $\alpha = 15^h07^m10^s11$ and $\delta = -17°32'13''3$

19

Rectangular Coordinates of the Sun

The rectangular geocentric equatorial coordinates X, Y, Z of the Sun are needed for the calculation of an ephemeris of a minor planet or a comet (see Chapters 25 and 26). The origin of these coordinates is the center of the Earth. The X-axis is directed towards the vernal equinox (longitude 0°); the Y-axis lies in the plane of the equator too and is directed towards longitude 90°, while the Z-axis is directed towards the north celestial pole.

The values of X, Y, Z are given for each day at 0^h ET in the *Astronomical Ephemeris* ; they are expressed in astronomical units. If the *A.E.* is not available, or for an instant in the past or in the future, the Sun's rectangular geocentric equatorial coordinates can be calculated from

$$X = R \cos \odot$$
$$Y = R \sin \odot \cos \varepsilon \qquad\qquad (19.1)$$
$$Z = R \sin \odot \sin \varepsilon$$

where R is the Sun's radius vector expressed in astronomical units, $\odot$ the Sun's true longitude referred to the mean equinox of date, and ε the (mean) obliquity of the ecliptic for that date. The quantities R and $\odot$ can be calculated by the method given in Chapter 18, while ε is given by formula (18.4).

In the formulae (19.1) the latitude of the Sun, which always is very small, has been neglected.

However, the coordinates X, Y, Z, calculated as explained above, are referred to the mean equator and mean equinox of the date. In most cases, it will be necessary to have these coordinates referred to another equator and equinox, for example for the standard equinox of 1950.0. This can be performed in the following way.

If X_0 , Y_0 , Z_0 are the values at the initial equinox, and X, Y, Z the values at the final equinox, then

$$X = X_x X_0 + Y_x Y_0 + Z_x Z_0$$

$$Y = X_y X_0 + Y_y Y_0 + Z_y Z_0 \qquad\qquad (19.2)$$

$$Z = X_z X_0 + Y_z Y_0 + Z_z Z_0$$

where

$$X_x = \cos \zeta \cos z \cos \theta - \sin \zeta \sin z$$

$$X_y = \sin \zeta \cos z + \cos \zeta \sin z \cos \theta$$

$$X_z = \cos \zeta \sin \theta$$

$$Y_x = -\cos \zeta \sin z - \sin \zeta \cos z \cos \theta$$

$$Y_y = \cos \zeta \cos z - \sin \zeta \sin z \cos \theta$$

$$Y_z = -\sin \zeta \sin \theta$$

$$Z_x = -\cos z \sin \theta$$

$$Z_y = -\sin z \sin \theta$$

$$Z_z = \cos \theta$$

the values of ζ, z and θ being found by the formulae (14.2).

It may be interesting to note that we have, approximately,

$$Y_x = -X_y \qquad\qquad Z_x = -X_z \qquad\qquad Z_y = Y_z$$

Example 19.a : Find the X, Y, Z coordinates of the Sun, referred to the equator and ecliptic of 1950.0, for 1978 November 12 at $0\,h$ ET = JD 2443 824.5.

In Example 18.a, we have found the following values for that instant :

$$\odot = 229°\!.25049 \qquad\qquad R = 0.98984$$

Formulae (18.1) and (18.4) give

$$T = +0.788\ 624\ 230 \qquad\qquad \varepsilon = 23°\!.442\ 031$$

Then formulae (19.1) give

$$X = -0.646\ 121$$
$$Y = -0.687\ 981$$
$$Z = -0.298\ 316$$

These values are referred to the equator and ecliptic of the date. They must be reduced to those of 1950.0 by means of formulae (19.2) but firstly we have to calculate ζ, z and θ (Chapter 14). We find

$$\tau_0 = \frac{2443\ 824.5 - 2415\ 020.313}{36524.2199} = +0.788\ 632\ 504$$

$$\tau = \frac{2433\ 282.423 - 2443\ 824.5}{36524.2199} = -0.288\ 632\ 503$$

$$\zeta = -665\overset{''}{.}374 = -0\overset{\circ}{.}184\ 826$$

$$z = -665\overset{''}{.}309 = -0\overset{\circ}{.}184\ 808$$

$$\theta = -578\overset{''}{.}457 = -0\overset{\circ}{.}160\ 682$$

Then,

$X_x = +0.999\ 9753$	$Y_x = +0.006\ 4513$	$Z_x = +0.002\ 8044$
$X_y = -0.006\ 4513$	$Y_y = +0.999\ 9792$	$Z_y = -0.000\ 0090$
$X_z = -0.002\ 8044$	$Y_z = -0.000\ 0090$	$Z_z = +0.999\ 9961$

and finally, by formulae (19.2),

$$X_{1950} = -0.651\ 38$$
$$Y_{1950} = -0.683\ 80$$
$$Z_{1950} = -0.296\ 50$$

According to the $A.E.$, the correct values are

$$-0.651\ 3639$$
$$-0.683\ 8057$$
$$-0.296\ 5014$$

20

Equinoxes and Solstices

The times of the equinoxes and solstices are the instants when the apparent longitude of the Sun is a multiple of 90 degrees. These instants can be calculated as follows.

Firstly, find an approximate time (in Julian Days) by means of the formula

$$JD = (\text{year} + k/4) \times 365.2422 + 1721\,141.3 \qquad (20.1)$$

where "year" is an integer, and

$k = 0$ for the March equinox,
 1 for the June solstice,
 2 for the September equinox,
 3 for the December solstice.

For the JD given by formula (20.1), calculate the Sun's apparent longitude Θ_{app} by the method described in Chapter 18. A correction to the JD is then given by

$$+ 58 \sin (k.90° - \Theta_{app}) \quad \text{days} \qquad (20.2)$$

Using the new value for JD, the calculation should be repeated if necessary, until one finds a correction that is small, say less than 0.001 day.

The final JD can be converted into ordinary calendar date by means of the method described in Chapter 3. The result is expressed in Ephemeris Time.

Example 20.a : Find the instant of the September equinox of the year 1979.

Putting year = +1979, and $k = 2$, in formula (20.1), we find the approximate value JD = 2444\,138.24.

For this instant we find, by the method described in Chapter 18, Θ_{app} = 28978$\overset{\circ}{.}$144 = 178$\overset{\circ}{.}$144 and the correction, given by formula (20.2), is then

$$+ 58 \; \sin \, (180° \, - \, 178\overset{\circ}{.}144) \, = \, + 1.88 \;\; day.$$

The corrected instant is thus

$$JD \; = \; 2444\,138.24 \, + \, 1.88 \, = \; 2444\,140.12.$$

With this new value, one finds Θ_{app} = 179$\overset{\circ}{.}$983, and the new correction is +0.017 day, giving the new corrected value for the instant JD = 2444 140.137.

Using this latter value again, one finds Θ_{app} = 180$\overset{\circ}{.}$000, which shows that the correct instant is indeed JD = 2444 140.137. This corresponds to 1979 September 23 at $15^h17^m17^s$ ET, which must be rounded to 15^h16^m UT.

(In 1979, the difference ET – UT is approximately 50 seconds). The correct value, as given by the $A.E.$, is 15^h17^m UT.

Instead of formula (20.1), better approximate times can be obtained as follows.

March Equinox :

$$JD \; = \; 1721\,139.2855 \, + \, 365.242\,1376 \; Y \, + \, 0.067\,9190 \; y^2 \, - \, 0.002\,7879 \; y^3$$

June Solstice :

$$JD \; = \; 1721\,233.2486 \, + \, 365.241\,7284 \; Y \, - \, 0.053\,0180 \; y^2 \, + \, 0.009\,3320 \; y^3$$

September Equinox :

$$JD \; = \; 1721\,325.6978 \, + \, 365.242\,5055 \; Y \, - \, 0.126\,6890 \; y^2 \, + \, 0.001\,9401 \; y^3$$

December Solstice :

$$JD \; = \; 1721\,414.3920 \, + \, 365.242\,8898 \; Y \, - \, 0.010\,9650 \; y^2 \, - \, 0.008\,4885 \; y^3$$

In these formulae, Y is the year, and $y = Y/1000$. It is important to note that Y must be an *integer*. Other values for Y will give meaningless results !

The times obtained by means of these formulae will generally be not more than 15 minutes in error.

21

EQUATION OF TIME

The equation of time is the difference between the right ascensions of the apparent (true) Sun and the fictitious mean Sun. If the *A.E.* is available, the equation of time E at 0^h UT can be calculated from

E = 12 hours + apparent sidereal time at 0^h UT
               − apparent right ascension of Sun at 0^h ET
               − 0.002738 ΔT

where ΔT = ET − UT.

Example 21.a : Calculate the equation of time on 1978 January 21 at 0^h Universal Time.

From the *Astronomical Ephemeris* we take the following values :

apparent sidereal time at 0^h UT = $8^h00^m01^s.193$
apparent right ascension of Sun at 0^h ET = $20^h11^m10^s.78$
ΔT = +48.6 seconds

Hence,

$E = 20^h00^m01^s.193 - 20^h11^m10^s.78 - (0.002738 \times 48^s.6)$
   $= -11^m09^s.72$

If the *A.E.* is not available, the equation of time at any instant can be calculated by means of the following formula given by W.M. Smart (*Text-Book on Spherical Astronomy*, page 149 of the edition of 1956) :

$$E = y \sin 2L - 2e \sin M + 4e\,y \sin M \cos 2L$$
$$- \frac{1}{2} y^2 \sin 4L - \frac{5}{4} e^2 \sin 2M \tag{21.1}$$

where $y = \tan^2\frac{\varepsilon}{2}$, ε being the obliquity of the ecliptic,

  L = Sun's mean longitude,
  e = eccentricity of the Earth's orbit,
  M = Sun's mean anomaly.

The values of ε, L, e and M can be found by means of the formulae given in Chapter 18.

The value of E given by formula (21.1) is expressed in radians. The result may be converted into degrees, and then into hours and decimals by division by 15.

Example 21.b : Calculate the equation of time on 1978 January 21 at 0^h ET = JD 2443 529.5.

We find successively

$$T = +0.780\ 547\ 5702$$
$$L = 28380\overset{\circ}{.}00957 = 300\overset{\circ}{.}00957$$
$$M = 28457\overset{\circ}{.}44655 = 17\overset{\circ}{.}44655$$
$$e = 0.016\ 718\ 34$$
$$\varepsilon = 23\overset{\circ}{.}442\ 136$$
$$y = 0.043\ 045\ 274$$

Formula (21.1) then gives $E = -0.048\ 743\ 490$ radian
$$= -2\overset{\circ}{.}792\ 7963$$
$$= -11 \text{ minutes } 10.3 \text{ seconds}$$

22

EQUATION OF KEPLER

The equation of Kepler is

$$E = M + e \sin E \qquad (22.1)$$

where e is the eccentricity of the planet's orbit, M the planet's mean anomaly at a given instant, and E the eccentric anomaly. Generally, e and M are given, and the equation must be solved for E, as in Chapters 18, 25 and 39. The eccentric anomaly E is an auxiliary quantity which is needed to find the true anomaly v.

Equation (22.1) is a transcendental equation in E and cannot be solved directly. We will describe two iteration methods for finding E (iteration = repetition), and finally a formula which gives an approximate result.

FIRST METHOD

It should be noted that in formula (22.1) the angles M and E should be expressed in *radians*. On the calculating machine, the calculations must thus be performed in "radian mode". This can be avoided by multiplying e by $180/\pi$ (conversion from radians into degrees) in equation (22.1). Let e_o be the thus "modified" eccentricity. Kepler's equation is then

$$E = M + e_o \sin E \qquad (22.2)$$

and now we can calculate with ordinary degrees.

To solve equation (22.2), give an approximate value to E in the right side of the formula. Then the formula will give a better approximation for E. This is repeated until the required accuracy is obtained ; this process can be performed automatically on a programmable calculator. For the first approximation, use $E = M$.

We thus have

$$E_0 = M$$
$$E_1 = M + e \sin E_0$$
$$E_2 = M + e \sin E_1$$
$$E_3 = M + e \sin E_2$$
$$\text{etc.}$$

E_1, E_2, E_3, etc. are successive and better approximations for the eccentric anomaly E.

Example 22.a : Solve the equation of Kepler for $e = 0.100$ and $M = 5°$, to an accuracy of 0.000 001 degree.

We find
$$e_o = 0.100 \times 180/\pi = 5°729\,577\,95,$$

and the equation of Kepler becomes
$$E = 5 + 5.729\,577\,95 \sin E$$

where all quantities are in degrees. Starting with $E = M = 5°$, we obtain successively :

$$5.000\,000$$
$$5.499\,366$$
$$5.549\,093$$
$$5.554\,042$$
$$5.554\,535$$
$$5.554\,584$$
$$5.554\,589$$
$$5.554\,589$$

Hence, the required value is $E = 5°554\,589$.

Second Method

The first method is very simple, and there will be no problems when e is small. However, the number of required iterations is generally increasing with e. For example, for $e = 0.990$ and $M = 2°$ the successive values of the iteration procedure are as follows :

2.000 000	15.168 909	24.924 579	29.813 009
3.979 598	16.842 404	25.904 408	30.200 940
5.936 635	18.434 883	26.780 556	30.533 515
7.866 758	19.937 269	27.557 863	30.817 592
9.763 644	21.341 978	28.242 483	.
11.619 294	22.643 349	28.841 471	:
13.424 417	23.837 929	29.362 399	:

After the 50th iteration, the result $(32°345\,452)$ still differs from the correct value $(32.361\,007)$ by more than 0.01 degree.

When e is larger than 0.4 or 0.5, the convergence may be so slow that a better iteration formula should be used : a better value E_1 for E is

$$E_1 = E_0 + \frac{M + e_o \sin E_0 - E_0}{1 - e \cos E_0} \qquad (22.3)$$

where E_0 is the lastly obtained value for E. In this formula, all quantities are expressed in degrees. It is important to note that the numerator of the fraction contains the "modified" eccentricity e_o defined before, while the denominator contains the ordinary eccentricity e.

Here, again, the process can be repeated as often as is necessary.

Example 22.b : Same problem as in Example 22.a, but now using formula (22.3).

In this case, formula (22.3) takes the following form :

$$E_1 = E_0 + \frac{5 + 5.729\,577\,95 \sin E_0 - E_0}{1 - 0.100 \cos E_0}$$

Starting with $E_0 = M = 5°$, we obtain the following values :

E_0	*correction*	E_1
5.000 000 000	+0.554 616 193	5.554 616 193
5.554 616 193	−0.000 026 939	5.554 589 254
5.554 589 254	−0.000 000 001	5.554 589 253

In this case, an accuracy of 0.000 000 001 degree is obtained after only three iterations.

As an exercise, try the second method on the case mentioned before : $e = 0.99$, $M = 2°$. After only nine or ten iterations, an accuracy of 0.000 0001 degree is reached.

In the first as well as in the second method, a test must be included in the program, because a new iteration should be performed only as long as the required accuracy (for instance 0.000 001 degree) has not been reached. It is important to note a difference in the test for the two methods.

In the first method, formula (22.2) gives directly a new approximation for E. This new value, after being stored, must be compared to the previous one, which thus should be temporarily retained in the machine. Thus, this method requires the use of two registers, one containing the new value of E, and the other containing the previous value.

In the second method, formula (22.3) too gives a new approximation E_1 for the eccentric anomaly, but the fraction in the second member is actually a *correction* to the previous value E_0. On many machines, this correction can be added directly to the value of E_0 contained in a register ("storage register arithmetic", for instance the instruction STO + 0 on the HP-67 machine), after which the absolute value of the correction (which is still displayed) can be tested. This procedure requires only one register for the eccentric anomaly.

Third Method

The formula

$$\tan E = \frac{\sin M}{\cos M - e} \tag{22.4}$$

gives an *approximate* value for E, and is valid only for small values of the eccentricity.

For the same data as in Example 22.a, the formula (22.4) gives

$$\tan E = \frac{+0.087\ 1557}{+0.896\ 1947}$$

whence $E = 5°554\ 599$, the exact value being $5°554\ 589$, an error of only $0''035$. (But for the same eccentricity and $M = 82°$, the error amounts to $35''$).

The greatest error due to the use of formula (22.4) is

$$
\begin{array}{ll}
0°0327 & \text{for } e = 0.15 \\
0.0783 & \text{for } e = 0.20 \\
0.1552 & \text{for } e = 0.25 \\
1.42 & \text{for } e = 0.50 \\
24.7 & \text{for } e = 0.99
\end{array}
$$

For the orbit of the Earth ($e = 0.01674$), the error will be less than $0''2$. In that case, formula (22.4) can safely be used except when very high accuracy is needed.

23

Elements of the Planetary Orbits

The orbital elements of the major planets can be expressed as polynomials of the form

$$a_0 + a_1 T + a_2 T^2 + a_3 T^3$$

where T is the time measured in Julian centuries of 36525 ephemeris days from the epoch 1900 January 0.5 ET = JD 2415 020.0. In other words,

$$T = \frac{JD - 2415\,020.0}{36525} \tag{23.1}$$

This quantity is negative before the beginning of the year 1900, positive afterwards. The orbital elements are :

L = mean longitude of the planet ;

a = semimajor axis of the orbit (in fact, this elements is a constant for each planet) ;

e = eccentricity of the orbit ;

i = inclination on the plane of the ecliptic ;

ω = argument of perihelion ;

Ω = longitude of ascending node.

The longitude of the perihelion can be calculated from $\pi = \omega + \Omega$, and the planet's mean anomaly is

$$M = L - \pi = L - \omega - \Omega$$

See also Chapter 25 for the mean anomalies.

The perihelion distance q and the aphelion distance Q are

$$q = a\,(1 - e) \qquad\qquad Q = a\,(1 + e)$$

We have $q + Q = 2a$.

The quantities L and π are measured in two different planes, namely from the vernal equinox along the ecliptic to the orbit's ascending node , and then from this node along the orbit.

Table 23.A gives the coefficients a_i for the orbital elements of the planets Mercury to Neptune. The values for Mercury and Venus are those given by S. Newcomb. The values for Mars are due to F.E. Ross. The elements for Jupiter, Saturn, Uranus and Neptune, due to Gaillot, are *not* affected by the periodic terms of short and long period ; thus they correspond to the purely secular terms.

The elements for the Earth are not given in Table 23.A. Since for this planet we have $i = 0$, the angles ω and Ω are not determined. The Earth's mean anomaly and orbital eccentricity are equal to those of the Sun (see Chapter 18), while the mean longitude and the longitude of the perihelion of the Earth are equal to those of the Sun increased by 180 degrees. Finally, for the Earth we have $a = 1.000\,0002$.

In Table 23.A, the values for the angular quantities L, i, ω and Ω are expressed in degrees and decimals.

Example 23.a :   Calculate the orbital elements of Mercury on 1978 June 24.0 ET.

We have (see Chapter 3)

$$1978 \text{ June } 24.0 = \text{JD } 2443\,683.5$$

whence, by formula (23.1),

$$T = +0.784\,763\,8604$$

Consequently, from Table 23.A, we find :

$$L = 178°179\,078 + (149\,474°070\,78 \times 0.784\,763\,8604)$$
$$+ (0.000\,3011)(0.784\,763\,8604)^2$$
$$= 117\,480°0281 = 120°0281$$

$$a = 0.387\,0986$$

$$e = 0.205\,630\,25$$

$$i = 7°004\,330$$

$$\omega = 29°044\,410$$

$$\Omega = 48°076\,160 \qquad M = 42°9075$$

TABLE 23.A

Elements for the mean equinox of the date

	a_0	a_1	a_2	a_3
MERCURY				
L	178.179 078	+ 149 474.070 78	+ 0.000 3011	
a	0.387 0986			
e	0.205 614 21	+ 0.000 020 46	− 0.000 000 030	
i	7.002 881	+ 0.001 8608	− 0.000 0183	
ω	28.753 753	+ 0.370 2806	+ 0.000 1208	
Ω	47.145 944	+ 1.185 2083	+ 0.000 1739	
VENUS				
L	342.767 053	+ 58 519.211 91	+ 0.000 3097	
a	0.723 3316			
e	0.006 820 69	− 0.000 047 74	+ 0.000 000 091	
i	3.393 631	+ 0.001 0058	− 0.000 0010	
ω	54.384 186	+ 0.508 1861	− 0.001 3864	
Ω	75.779 647	+ 0.899 8500	+ 0.000 4100	
MARS				
L	293.737 334	+ 19 141.695 51	+ 0.000 3107	
a	1.523 6883			
e	0.093 312 90	+ 0.000 092 064	− 0.000 000 077	
i	1.850 333	− 0.000 6750	+ 0.000 0126	
ω	285.431 761	+1.069 7667	+ 0.000 1313	+ 0.000 004 14
Ω	48.786 442	+ 0.770 9917	− 0.000 0014	− 0.000 005 33

TABLE 23.A (continuation)

	a_0	a_1	a_2	a_3
JUPITER				
L	238.049 257	+ 3036.301 986	+ 0.000 3347	− 0.000 001 65
a	5.202 561			
e	0.048 334 75	+ 0.000 164 180	− 0.000 000 4676	− 0.000 000 0017
i	1.308 736	− 0.005 6961	+ 0.000 0039	
ω	273.277 558	+ 0.599 4317	+ 0.000 704 05	+ 0.000 005 08
Ω	99.443 414	+ 1.010 5300	+ 0.000 352 22	− 0.000 008 51
SATURN				
L	266.564 377	+ 1223.509 884	+ 0.000 3245	− 0.000 0058
a	9.554 747			
e	0.055 892 32	− 0.000 345 50	− 0.000 000 728	+ 0.000 000 000 74
i	2.492 519	− 0.003 9189	− 0.000 015 49	+ 0.000 000 04
ω	338.307 800	+ 1.085 2207	+ 0.000 978 54	+ 0.000 009 92
Ω	112.790 414	+ 0.873 1951	− 0.000 152 18	− 0.000 005 31
URANUS				
L	244.197 470	+ 429.863 546	+ 0.000 3160	− 0.000 000 60
a	19.218 14			
e	0.046 3444	− 0.000 026 58	+ 0.000 000 077	
i	0.772 464	+ 0.000 6253	+ 0.000 0395	
ω	98.071 581	+ 0.985 7650	− 0.001 0745	− 0.000 000 61
Ω	73.477 111	+ 0.498 6678	+ 0.001 3117	

TABLE 23.A   (end)

	a_0	a_1	a_2	a_3
NEPTUNE				
L	84.457 994	+ 219.885 914	+ 0.000 3205	− 0.000 000 60
a	30.109 57			
e	0.008 997 04	+ 0.000 006 330	− 0.000 000 002	
i	1.779 242	− 0.009 5436	− 0.000 0091	
ω	276.045 975	+ 0.325 6394	+ 0.000 140 95	+ 0.000 004 113
Ω	130.681 389	+ 1.098 9350	+ 0.000 249 87	− 0.000 004 718

The elements calculated by means of the coefficients of Table 23.A are referred to the mean equinox of the date, that is to the ecliptic of the date and to the mean equator of the date. Consequently, those coefficients should be used if one wishes to calculate planetary positions referred to the mean equinox of the date.

In some cases, however, it may be desirable to refer the elements i, ω, Ω to a standard equinox. This is the case, for instance, when one wishes to calculate the least distance between the orbit of a comet and that of a major planet, when the elements of the first orbit are referred to a standard equinox.

By means of the formulae of Chapter 17, it is possible to convert the elements i, ω, Ω from one equinox to another one. However, by means of Tables 23.B and 23.C it is possible to calculate these elements for the major planets directly, referred to the standard equinox of either 1950.0 or 2000.0. The corresponding dates are

$$1950.0 \;=\; 1950\ \text{January}\ 0.923 \;=\; \text{JD}\ 2433\,282.423$$

$$2000.0 \;=\; 2000\ \text{January}\ 1.5 \;\;\;\; =\; \text{JD}\ 2451\,545.0$$

It should be noted that, while 1950.0 corresponds to the beginning of the Besselian year 1950 and is 50 *tropical* years later than the epoch 1900.0 = 1900 January 0.813 ET = JD 2415 020.313, the new standard epoch, designated 2000.0, will be exactly 36525 days after the epoch JD 2415 020.0 = 1900 January 0.5.

TABLE   23.B

Elements for equinox 1950.0

	a_0	a_1	a_2	a_3
MERCURY				
i	7.006 790	− 0.005 9671	+ 0.000 000 70	− 0.000 000 036
ω	28.796 761	+ 0.284 3099	+ 0.000 074 64	+ 0.000 000 043
Ω	47.801 352	− 0.125 5041	− 0.000 088 63	− 0.000 000 068
VENUS				
i	3.394 552	− 0.000 8226	− 0.000 032 51	+ 0.000 000 018
ω	54.493 527	+ 0.289 3249	− 0.001 144 35	− 0.000 000 792
Ω	76.368 593	− 0.277 7139	− 0.000 140 39	+ 0.000 000 767
EARTH				
i	− 0.006 540	+ 0.013 0855	− 0.000 009 33	+ 0.000 000 014
ω	287.390 758	+ 0.564 7073	+ 0.000 136 10	+ 0.000 003 333
Ω	174.528 170	− 0.241 5735	+ 0.000 007 94	− 0.000 000 028
MARS				
i	1.854 113	− 0.008 1839	− 0.000 023 05	− 0.000 000 045
ω	285.597 172	+ 0.738 5934	+ 0.000 466 47	+ 0.000 006 962
Ω	49.319 212	− 0.294 0497	− 0.000 644 35	− 0.000 008 182
JUPITER				
i	1.307 028	− 0.002 2192	+ 0.000 029 52	+ 0.000 000 125
ω	273.553 214	+ 0.047 5910	− 0.000 210 41	+ 0.000 009 039
Ω	99.865 881	+ 0.166 1852	+ 0.000 958 57	− 0.000 012 500

TABLE 23.B (end)

	a_0	a_1	a_2	a_3
SATURN				
i	2.489 374	+ 0.002 4190	− 0.000 050 22	+ 0.000 000 002
ω	338.439 665	+ 0.821 8494	+ 0.000 706 12	+ 0.000 006 174
Ω	113.356 715	− 0.259 7237	− 0.000 188 62	− 0.000 001 587
URANUS				
i	0.773 723	− 0.001 7599	− 0.000 000 22	+ 0.000 000 121
ω	98.546 561	+ 0.032 5540	− 0.000 501 25	+ 0.000 013 998
Ω	73.700 227	+ 0.055 7505	+ 0.000 429 88	− 0.000 014 630
NEPTUNE				
i	1.774 485	− 0.000 0150	− 0.000 002 27	+ 0.000 000 018
ω	276.190 852	+ 0.036 7891	+ 0.000 038 42	+ 0.000 002 218
Ω	131.234 637	− 0.008 3952	+ 0.000 044 35	− 0.000 002 849

In the case of the Earth, if the inclination is found to be negative, then ω and Ω should *both* be increased or decreased by 180 degrees.

TABLE 23.C

Elements for equinox 2000.0

	a_0	a_1	a_2	a_3
MERCURY				
i	7.010 678	$-$ 0.005 9556	$+$ 0.000 000 69	$-$ 0.000 000 035
ω	28.839 814	$+$ 0.284 2765	$+$ 0.000 074 45	$+$ 0.000 000 043
Ω	48.456 876	$-$ 0.125 4715	$-$ 0.000 088 44	$-$ 0.000 000 068
VENUS				
i	3.395 459	$-$ 0.000 7913	$-$ 0.000 032 50	$+$ 0.000 000 018
ω	54.602 827	$+$ 0.289 2764	$-$ 0.001 144 64	$-$ 0.000 000 794
Ω	76.957 740	$-$ 0.277 6656	$-$ 0.000 140 10	$+$ 0.000 000 769
EARTH				
i	$-$0.013 0762	$+$ 0.013 0855	$-$ 0.000 009 27	$+$ 0.000 000 014
ω	287.511 505	$+$ 0.564 7920	$+$ 0.000 136 10	$+$ 0.000 003 333
Ω	175.105 679	$-$ 0.241 6582	$+$0.000 007 94	$-$ 0.000 000 028
MARS				
i	1.857 866	$-$ 0.008 1565	$-$ 0.000 023 04	$-$ 0.000 000 044
ω	285.762 379	$+$ 0.738 7251	$+$ 0.000 465 56	$+$ 0.000 006 939
Ω	49.852 347	$-$ 0.294 1821	$-$ 0.000 643 44	$-$ 0.000 008 159
JUPITER				
i	1.305 288	$-$ 0.002 2374	$+$ 0.000 029 42	$+$ 0.000 000 127
ω	273.829 584	$+$ 0.047 8404	$-$ 0.000 218 57	$+$ 0.000 008 999
Ω	100.287 838	$+$ 0.165 9357	$+$ 0.000 966 72	$-$ 0.000 012 460

TABLE 23.C (end)

	a_0	a_1	a_2	a_3
SATURN				
i	2.486 204	+ 0.002 4449	− 0.000 050 17	+ 0.000 000 002
ω	338.571 353	+ 0.822 0515	+ 0.000 707 47	+ 0.000 006 177
Ω	113.923 406	− 0.259 9254	− 0.000 189 97	− 0.000 001 589
URANUS				
i	0.774 950	− 0.001 7660	− 0.000 000 27	+ 0.000 000 123
ω	99.021 587	+ 0.033 7219	− 0.000 498 12	+ 0.000 013 904
Ω	73.923 501	+ 0.054 5828	+ 0.000 426 74	− 0.000 014 536
NEPTUNE				
i	1.769 715	− 0.000 0144	− 0.000 002 27	+ 0.000 000 018
ω	276.335 328	+ 0.036 8127	+ 0.000 038 49	+ 0.000 002 226
Ω	131.788 486	− 0.008 4187	+ 0.000 044 28	− 0.000 002 858

We see that the inclination of Mercury's orbit on the ecliptic of the date is increasing, but that it is decreasing with respect to the fixed ecliptic of either 1950.0 or 2000.0. The opposite occurs for Saturn.

Between $T = -20$ and $T = +20$, Venus' orbital inclination on the ecliptic of the date is continuously increasing, but with respect to the fixed ecliptic of 1950.0 Venus' inclination reached a maximum about the year +650.

Uranus' inclination on the ecliptic of the date reached a minimum about the year +1110, but with respect to the fixed equinoxes of 1950.0 and 2000.0 its value is continuously decreasing during the time period considered here.

Between $T = -20$ and $T = +20$, Neptune's orbital inclination on the ecliptic of the date is continuously decreasing, but with respect to the fixed ecliptic of 1950.0 Neptune's inclination reached a flat maximum about the year +1550.

The longitudes of the nodes, referred to the equinox of the date, are increasing for all planets. But with respect to the fixed equinoxes of 1950.0 and 2000.0 these longitudes are decreasing except for Jupiter and Uranus.

When the elements are taken from Table 23.B or 23.C, the mean longitude L referred to the same standard equinox can be found from

$$L = \Omega + \omega + M$$

where the mean anomaly M can be found either in Chapter 24 (for Uranus and Neptune), or in Chapter 25 (for the other planets).

24

PLANETS : PRINCIPAL PERTURBATIONS

In this Chapter we will mention the most important perturbations in the motion of the planets Mercury, Venus, Mars, Jupiter, Saturn, Uranus and Neptune. These periodic terms can be used if a better accuracy is needed than by using the data of Chapter 23 alone. The perturbations in the motions of the giant planets are particularly important ; in longitude, they can be larger than 0.3 degree for Jupiter, and larger than 1.0 degree for Saturn. For the Earth (Sun) the most important perturbations have been given in Chapter 18.

In the expressions given below, T is the time in Julian centuries from 1900 January 0.5 ET ; see formula (23.1).

M, the Sun's mean anomaly, can be calculated by means of the expression given on the first page of Chapter 18.

The mean anomalies of Mercury, Venus, Mars, Jupiter and Saturn are denoted by M_1 , M_2 , M_4 , M_5 and M_6 , and can be found by means of the formulae given in Chapter 25.

MERCURY

Perturbations in longitude

$+0°.00\ 204 \times \cos(5M_2 - 2M_1 + 12°.220)$
$+0.00\ 103 \quad \cos(2M_2 - M_1 - 160°.692)$
$+0.00\ 091 \quad \cos(2M_5 - M_1 - 37°.003)$
$+0.00\ 078 \quad \cos(5M_2 - 3M_1 + 10°.137)$

Perturbations in radius vector

$+0.000\ 007\ 525 \times \cos(2M_5 - M_1 + 53°.013)$
$+0.000\ 006\ 802 \quad \cos(5M_2 - 3M_1 - 259°.918)$
$+0.000\ 005\ 457 \quad \cos(2M_2 - 2M_1 - 71°.188)$
$+0.000\ 003\ 569 \quad \cos(5M_2 - M_1 - 77°.75)$

Term of long period in the *mean* longitude and in the mean anomaly :

$$+ 0\overset{\circ}{.}00077 \ \sin\,(237\overset{\circ}{.}24 + 150\overset{\circ}{.}27 \ T)$$

Perturbations in longitude

$+0\overset{\circ}{.}00\,313 \times \cos\,(2M - 2M_2 - 148\overset{\circ}{.}225)$
$+0.00\,198 \quad \cos\,(3M - 3M_2 + 2\overset{\circ}{.}565)$
$+0.00\,136 \quad \cos\,(M - M_2 - 119\overset{\circ}{.}107)$
$+0.00\,096 \quad \cos\,(3M - 2M_2 - 135\overset{\circ}{.}912)$
$+0.00\,082 \quad \cos\,(M_5 - M_2 - 208\overset{\circ}{.}087)$

Perturbations in radius vector

$+0.000\,022\,501 \times \cos\,(2M - 2M_2 - 58\overset{\circ}{.}208)$
$+0.000\,019\,045 \quad \cos\,(3M - 3M_2 + 92\overset{\circ}{.}577)$
$+0.000\,006\,887 \quad \cos\,(M_5 - M_2 - 118\overset{\circ}{.}090)$
$+0.000\,005\,172 \quad \cos\,(M - M_2 - 29\overset{\circ}{.}110)$
$+0.000\,003\,620 \quad \cos\,(5M - 4M_2 - 104\overset{\circ}{.}208)$
$+0.000\,003\,283 \quad \cos\,(4M - 4M_2 + 63\overset{\circ}{.}513)$
$+0.000\,003\,074 \quad \cos\,(2M_5 - 2M_2 - 55\overset{\circ}{.}167)$

The term of long period (with coefficient $0\overset{\circ}{.}00077$) should be added to both the mean longitude and mean anomaly *before* the equation of Kepler is solved. All other periodic terms must be added to the longitude and to the radius vector obtained *after* solving Kepler's equation.

MARS

Terms of long period in the *mean* longitude and in the mean anomaly :

$$-0\!\!\stackrel{\circ}{.}01\,133\ \sin\,(3M_5 - 8M_4 + 4M)$$
$$-0\!\!\stackrel{\circ}{.}00\,933\ \cos\,(3M_5 - 8M_4 + 4M)$$

Perturbations in longitude

$$
\begin{aligned}
&+0\!\!\stackrel{\circ}{.}00\,705 \times \cos\,(M_5 - M_4 - 48\!\!\stackrel{\circ}{.}958)\\
&+0.00\,607 \quad\ \ \cos\,(2M_5 - M_4 - 188\!\!\stackrel{\circ}{.}350)\\
&+0.00\,445 \quad\ \ \cos\,(2M_5 - 2M_4 - 191\!\!\stackrel{\circ}{.}897)\\
&+0.00\,388 \quad\ \ \cos\,(M - 2M_4 + 20\!\!\stackrel{\circ}{.}495)\\
&+0.00\,238 \quad\ \ \cos\,(M - M_4 + 35\!\!\stackrel{\circ}{.}097)\\
&+0.00\,204 \quad\ \ \cos\,(2M - 3M_4 + 158\!\!\stackrel{\circ}{.}638)\\
&+0.00\,177 \quad\ \ \cos\,(3M_4 - M_2 - 57\!\!\stackrel{\circ}{.}602)\\
&+0.00\,136 \quad\ \ \cos\,(2M - 4M_4 + 154\!\!\stackrel{\circ}{.}093)\\
&+0.00\,104 \quad\ \ \cos\,(M_5 + 17\!\!\stackrel{\circ}{.}618)
\end{aligned}
$$

Perturbations in radius vector

$$
\begin{aligned}
&+0.000\,053\,227 \times \cos\,(M_5 - M_4 + 41\!\!\stackrel{\circ}{.}1306)\\
&+0.000\,050\,989 \quad\ \ \cos\,(2M_5 - 2M_4 - 101\!\!\stackrel{\circ}{.}9847)\\
&+0.000\,038\,278 \quad\ \ \cos\,(2M_5 - M_4 - 98\!\!\stackrel{\circ}{.}3292)\\
&+0.000\,015\,996 \quad\ \ \cos\,(M - M_4 - 55\!\!\stackrel{\circ}{.}555)\\
&+0.000\,014\,764 \quad\ \ \cos\,(2M - 3M_4 + 68\!\!\stackrel{\circ}{.}622)\\
&+0.000\,008\,966 \quad\ \ \cos\,(M_5 - 2M_4 + 43\!\!\stackrel{\circ}{.}615)\\
&+0.000\,007\,914 \quad\ \ \cos\,(3M_5 - 2M_4 - 139\!\!\stackrel{\circ}{.}737)\\
&+0.000\,007\,004 \quad\ \ \cos\,(2M_5 - 3M_4 - 102\!\!\stackrel{\circ}{.}888)\\
&+0.000\,006\,620 \quad\ \ \cos\,(M - 2M_4 + 113\!\!\stackrel{\circ}{.}202)\\
&+0.000\,004\,930 \quad\ \ \cos\,(3M_5 - 3M_4 - 76\!\!\stackrel{\circ}{.}243)\\
&+0.000\,004\,693 \quad\ \ \cos\,(3M - 5M_4 + 190\!\!\stackrel{\circ}{.}603)\\
&+0.000\,004\,571 \quad\ \ \cos\,(2M - 4M_4 + 244\!\!\stackrel{\circ}{.}702)\\
&+0.000\,004\,409 \quad\ \ \cos\,(3M_5 - M_4 - 115\!\!\stackrel{\circ}{.}828)
\end{aligned}
$$

The terms of long period should be added to both the mean longitude and mean anomaly *before* the equation of Kepler is solved. All other periodic terms must be added to the longitude and to the radius vector obtained *after* solving Kepler's equation.

JUPITER

$$\upsilon = \frac{T}{5} + 0.1$$

$$P = 237°\!47555 + 3034°\!9061\,T$$

$$Q = 265°\!91650 + 1222°\!1139\,T$$

$$S = 243°\!51721 + 428°\!4677\,T$$

$$V = 5Q - 2P$$

$$W = 2P - 6Q + 3S$$

$$\zeta = Q - P$$

Perturbations in the mean longitude (A)

$+(0°\!331\,364 - 0°\!010\,281\,\upsilon - 0°\!004\,692\,\upsilon^2)\ \sin V$
$+(0°\!003\,228 - 0°\!064\,436\,\upsilon + 0°\!002\,075\,\upsilon^2)\ \cos V$
$-(0°\!003\,083 + 0°\!000\,275\,\upsilon - 0°\!000\,489\,\upsilon^2)\ \sin 2V$
$+0°\!002\,472\ \sin W$
$+0°\!013\,619\ \sin \zeta$
$+0°\!018\,472\ \sin 2\zeta$
$+0°\!006\,717\ \sin 3\zeta$
$+0°\!002\,775\ \sin 4\zeta$
$+(0°\!007\,275 - 0°\!001\,253\,\upsilon)\ \sin \zeta \sin Q$
$+0°\!006\,417\ \sin 2\zeta \sin Q$
$+0°\!002\,439\ \sin 3\zeta \sin Q$
$-(0°\!033\,839 + 0°\!001\,125\,\upsilon)\ \cos \zeta \sin Q$
$-0°\!003\,767\ \cos 2\zeta \sin Q$
$-(0°\!035\,681 + 0°\!001\,208\,\upsilon)\ \sin \zeta \cos Q$
$-0°\!004\,261\ \sin 2\zeta \cos Q$
$+0°\!002\,178\ \cos Q$
$+(-0°\!006\,333 + 0°\!001\,161\,\upsilon)\ \cos \zeta \cos Q$
$-0°\!006\,675\ \cos 2\zeta \cos Q$
$-0°\!002\,664\ \cos 3\zeta \cos Q$
$-0°\!002\,572\ \sin \zeta \sin 2Q$
$-0°\!003\,567\ \sin 2\zeta \sin 2Q$
$+0°\!002\,094\ \cos \zeta \cos 2Q$
$+0°\!003\,342\ \cos 2\zeta \cos 2Q$

Perturbations in the eccentricity

(The coefficients are given in units of the seventh decimal)

$+(3606 + 130\,\upsilon - 43\,\upsilon^2)\ \sin V$
$+(1289 - 580\,\upsilon)\ \cos V$

$-6764 \sin \zeta \sin Q$
$-1110 \sin 2\zeta \sin Q$
$-224 \sin 3\zeta \sin Q$
$-204 \sin Q$
$+(1284 + 116\,\upsilon) \cos \zeta \sin Q$
$+188 \cos 2\zeta \sin Q$
$+(1460 + 130\,\upsilon) \sin \zeta \cos Q$
$+224 \sin 2\zeta \cos Q$
$-817 \cos Q$
$+6074 \cos \zeta \cos Q$
$+992 \cos 2\zeta \cos Q$
$+508 \cos 3\zeta \cos Q$
$+230 \cos 4\zeta \cos Q$
$+108 \cos 5\zeta \cos Q$
$-(956 + 73\,\upsilon) \sin \zeta \sin 2Q$
$+448 \sin 2\zeta \sin 2Q$
$+137 \sin 3\zeta \sin 2Q$
$+(-997 + 108\,\upsilon) \cos \zeta \sin 2Q$
$+480 \cos 2\zeta \sin 2Q$
$+148 \cos 3\zeta \sin 2Q$
$+(-956 + 99\,\upsilon) \sin \zeta \cos 2Q$
$+490 \sin 2\zeta \cos 2Q$
$+158 \sin 3\zeta \cos 2Q$
$+179 \cos 2Q$
$+(1024 + 75\,\upsilon) \cos \zeta \cos 2Q$
$-437 \cos 2\zeta \cos 2Q$
$-132 \cos 3\zeta \cos 2Q$

Perturbations in the perihelion (B)

$+(0\overset{s}{.}007\,192 - 0\overset{s}{.}003\,147\,\upsilon) \sin V$
$+(-0\overset{s}{.}020\,428 - 0\overset{s}{.}000\,675\,\upsilon + 0\overset{s}{.}000\,197\,\upsilon^2) \cos V$
$+(0\overset{s}{.}007\,269 + 0\overset{s}{.}000\,672\,\upsilon) \sin \zeta \sin Q$
$-0\overset{s}{.}004\,344 \sin Q$
$+0\overset{s}{.}034\,036 \cos \zeta \sin Q$
$+0\overset{s}{.}005\,614 \cos 2\zeta \sin Q$
$+0\overset{s}{.}002\,964 \cos 3\zeta \sin Q$
$+0\overset{s}{.}037\,761 \sin \zeta \cos Q$
$+0\overset{s}{.}006\,158 \sin 2\zeta \cos Q$
$-0\overset{s}{.}006\,603 \cos \zeta \cos Q$
$-0\overset{s}{.}005\,356 \sin \zeta \sin 2Q$
$+0\overset{s}{.}002\,722 \sin 2\zeta \sin 2Q$
$+0\overset{s}{.}004\,483 \cos \zeta \sin 2Q \qquad -0\overset{s}{.}002\,536 \sin 2\zeta \cos 2Q$
$-0\overset{s}{.}002\,642 \cos 2\zeta \sin 2Q \qquad +0\overset{s}{.}005\,547 \cos \zeta \cos 2Q$
$+0\overset{s}{.}004\,403 \sin \zeta \cos 2Q \qquad -0\overset{s}{.}002\,689 \cos 2\zeta \cos 2Q$

If A is the sum of the perturbations in the mean longitude, B the sum of the perturbations in the perihelion, and e the orbital eccentricity *not* corrected for the perturbations, then the correction to the mean anomaly is

$$A - \frac{B}{e}$$

Perturbations in the semimajor axis
(The coefficients are given in units of the sixth decimal)

$$-263 \cos V$$
$$+205 \cos \zeta$$
$$+693 \cos 2\zeta$$
$$+312 \cos 3\zeta$$
$$+147 \cos 4\zeta$$
$$+299 \sin \zeta \sin Q$$
$$+181 \cos 2\zeta \sin Q$$
$$+204 \sin 2\zeta \cos Q$$
$$+111 \sin 3\zeta \cos Q$$
$$-337 \cos \zeta \cos Q$$
$$-111 \cos 2\zeta \cos Q$$

IMPORTANT NOTE. — The perturbations to the mean longitude, to the mean anomaly, to the eccentricity and to the semimajor axis should be added to the mean elements *before* solving the equation of Kepler, etc.

Calculate υ, V, W, ζ, etc. as for Jupiter, and moreover $\psi = S - Q$.

Perturbations in the mean longitude (A)

$+(-0\overset{\text{s}}{.}814\,181 + 0\overset{\text{s}}{.}018\,150\,\upsilon + 0\overset{\text{s}}{.}016\,714\,\upsilon^2)\,\sin\,V$
$+(-0\overset{\text{s}}{.}010\,497 + 0\overset{\text{s}}{.}160\,906\,\upsilon - 0\overset{\text{s}}{.}004\,100\,\upsilon^2)\,\cos\,V$
$+0\overset{\text{s}}{.}007\,581\,\sin\,2V$
$-0\overset{\text{s}}{.}007\,986\,\sin\,W$
$-0\overset{\text{s}}{.}148\,811\,\sin\,\zeta$
$-0\overset{\text{s}}{.}040\,786\,\sin\,2\zeta$
$-0\overset{\text{s}}{.}015\,208\,\sin\,3\zeta$
$-0\overset{\text{s}}{.}006\,339\,\sin\,4\zeta$
$-0\overset{\text{s}}{.}006\,244\,\sin\,Q$
$+(0\overset{\text{s}}{.}008\,931 + 0\overset{\text{s}}{.}002\,728\,\upsilon)\,\sin\,\zeta\,\sin\,Q$
$-0\overset{\text{s}}{.}016\,500\,\sin\,2\zeta\,\sin\,Q$
$-0\overset{\text{s}}{.}005\,775\,\sin\,3\zeta\,\sin\,Q$
$+(0\overset{\text{s}}{.}081\,344 + 0\overset{\text{s}}{.}003\,206\,\upsilon)\,\cos\,\zeta\,\sin\,Q$
$+0\overset{\text{s}}{.}015\,019\,\cos\,2\zeta\,\sin\,Q$
$+(0\overset{\text{s}}{.}085\,581 + 0\overset{\text{s}}{.}002\,494\,\upsilon)\,\sin\,\zeta\,\cos\,Q$
$+(0\overset{\text{s}}{.}025\,328 - 0\overset{\text{s}}{.}003\,117\,\upsilon)\,\cos\,\zeta\,\cos\,Q$
$+0\overset{\text{s}}{.}014\,394\,\cos\,2\zeta\,\cos\,Q$
$+0\overset{\text{s}}{.}006\,319\,\cos\,3\zeta\,\cos\,Q$
$+0\overset{\text{s}}{.}006\,369\,\sin\,\zeta\,\sin\,2Q$
$+0\overset{\text{s}}{.}009\,156\,\sin\,2\zeta\,\sin\,2Q$
$+0\overset{\text{s}}{.}007\,525\,\sin\,3\psi\,\sin\,2Q$
$-0\overset{\text{s}}{.}005\,236\,\cos\,\zeta\,\cos\,2Q$
$-0\overset{\text{s}}{.}007\,736\,\cos\,2\zeta\,\cos\,2Q$
$-0\overset{\text{s}}{.}007\,528\,\cos\,3\psi\,\cos\,2Q$

Perturbations in the eccentricity

(The coefficients are given in units of the seventh decimal)

$+(-7927 + 2548\,\upsilon + 91\,\upsilon^2)\,\sin\,V$
$+(13381 + 1226\,\upsilon - 253\,\upsilon^2)\,\cos\,V$
$+(248 - 121\,\upsilon)\,\sin\,2V$
$-(305 + 91\,\upsilon)\,\cos\,2V$
$+412\,\sin\,2\zeta$
$+12415\,\sin\,Q$
$+(390 - 617\,\upsilon)\,\sin\,\zeta\,\sin\,Q$
$+(165 - 204\,\upsilon)\,\sin\,2\zeta\,\sin\,Q$
$+26599\,\cos\,\zeta\,\sin\,Q$
$-4687\,\cos\,2\zeta\,\sin\,Q$

$-1870 \cos 3\zeta \sin Q$
$-821 \cos 4\zeta \sin Q$
$-377 \cos 5\zeta \sin Q$
$+497 \cos 2\psi \sin Q$
$+(163 - 611\,\upsilon) \cos Q$
$-12696 \sin \zeta \cos Q$
$-4200 \sin 2\zeta \cos Q$
$-1503 \sin 3\zeta \cos Q$
$-619 \sin 4\zeta \cos Q$
$-268 \sin 5\zeta \cos Q$
$-(282 + 1306\,\upsilon) \cos \zeta \cos Q$
$+(-86 + 230\,\upsilon) \cos 2\zeta \cos Q$
$+461 \sin 2\psi \cos Q$
$-350 \sin 2Q$
$+(2211 - 286\,\upsilon) \sin \zeta \sin 2Q$
$-2208 \sin 2\zeta \sin 2Q$
$-568 \sin 3\zeta \sin 2Q$
$-346 \sin 4\zeta \sin 2Q$
$-(2780 + 222\,\upsilon) \cos \zeta \sin 2Q$
$+(2022 + 263\,\upsilon) \cos 2\zeta \sin 2Q$
$+248 \cos 3\zeta \sin 2Q$
$+242 \sin 3\psi \sin 2Q$
$+467 \cos 3\psi \sin 2Q$
$-490 \cos 2Q$
$-(2842 + 279\,\upsilon) \sin \zeta \cos 2Q$
$+(128 + 226\,\upsilon) \sin 2\zeta \cos 2Q$
$+224 \sin 3\zeta \cos 2Q$
$+(-1594 + 282\,\upsilon) \cos \zeta \cos 2Q$
$+(2162 - 207\,\upsilon) \cos 2\zeta \cos 2Q$
$+561 \cos 3\zeta \cos 2Q$
$+343 \cos 4\zeta \cos 2Q$
$+469 \sin 3\psi \cos 2Q$
$-242 \cos 3\psi \cos 2Q$
$-205 \sin \zeta \sin 3Q$
$+262 \sin 3\zeta \sin 3Q$
$+208 \cos \zeta \cos 3Q$
$-271 \cos 3\zeta \cos 3Q$
$-382 \cos 3\zeta \sin 4Q$
$-376 \sin 3\zeta \cos 4Q$

Perturbations in the perihelion (B)

$+(0°077\,108 + 0°007\,186\,\upsilon - 0°001\,533\,\upsilon^2)\,\sin V$
$+(0°045\,803 - 0°014\,766\,\upsilon - 0°000\,536\,\upsilon^2)\,\cos V$
$-0°007\,075\,\sin \zeta$
$-0°075\,825\,\sin \zeta \sin Q$
$-0°024\,839\,\sin 2\zeta \sin Q$
$-0°008\,631\,\sin 3\zeta \sin Q$
$-0°072\,586\,\cos Q$
$-0°150\,383\,\cos \zeta \cos Q$
$+0°026\,897\,\cos 2\zeta \cos Q$
$+0°010\,053\,\cos 3\zeta \cos Q$
$-(0°013\,597 + 0°001\,719\,\upsilon)\,\sin \zeta \sin 2Q$
$+(-0°007\,742 + 0°001\,517\,\upsilon)\,\cos \zeta \sin 2Q$
$+(0°013\,586 - 0°001\,375\,\upsilon)\,\cos 2\zeta \sin 2Q$
$+(-0°013\,667 + 0°001\,239\,\upsilon)\,\sin \zeta \cos 2Q$
$+0°011\,981\,\sin 2\zeta \cos 2Q$
$+(0°014\,861 + 0°001\,136\,\upsilon)\,\cos \zeta \cos 2Q$
$-(0°013\,064 + 0°001\,628\,\upsilon)\,\cos 2\zeta \cos 2Q$

As for Jupiter, the correction to the mean longitude is A, and the correction to the mean anomaly is $A - \dfrac{B}{e}$.

Perturbations in the semimajor axis
(The coefficients are given in units of the sixth decimal)

$+572\,\upsilon \sin V$
$+2933\,\cos V$
$+33629\,\cos \zeta$
$-3081\,\cos 2\zeta$
$-1423\,\cos 3\zeta$
$-671\,\cos 4\zeta$
$-320\,\cos 5\zeta$
$+1098\,\sin Q$
$-2812\,\sin \zeta \sin Q$
$+688\,\sin 2\zeta \sin Q$
$-393\,\sin 3\zeta \sin Q$
$-228\,\sin 4\zeta \sin Q$
$+2138\,\cos \zeta \sin Q$
$-999\,\cos 2\zeta \sin Q$
$-642\,\cos 3\zeta \sin Q$
$-325\,\cos 4\zeta \sin Q$
$-890\,\cos Q$
$+2206\,\sin \zeta \cos Q$

$-1590\,\sin 2\zeta \cos Q$
$-647\,\sin 3\zeta \cos Q$
$-344\,\sin 4\zeta \cos Q$
$+2885\,\cos \zeta \cos Q$
$+(2172 + 102\,\upsilon)\,\cos 2\zeta \cos Q$
$+296\,\cos 3\zeta \cos Q$
$-267\,\sin 2\zeta \sin 2Q$
$-778\,\cos \zeta \sin 2Q$
$+495\,\cos 2\zeta \sin 2Q$
$+250\,\cos 3\zeta \sin 2Q$
$-856\,\sin \zeta \cos 2Q$
$+441\,\sin 2\zeta \cos 2Q$
$+296\,\cos 2\zeta \cos 2Q$
$+211\,\cos 3\zeta \cos 2Q$
$-427\,\sin \zeta \sin 3Q$
$+398\,\sin 3\zeta \sin 3Q$
$+344\,\cos \zeta \cos 3Q$
$-427\,\cos 3\zeta \cos 3Q$

Then, after the whole calculation (equation of Kepler, etc.), add the following perturbations to the heliocentric *latitude* :

$$+0\overset{\circ}{.}000\ 747\ \cos\ \zeta\ \sin\ Q$$
$$+0\overset{\circ}{.}001\ 069\ \cos\ \zeta\ \cos\ Q$$
$$+0\overset{\circ}{.}002\ 108\ \sin\ 2\zeta\ \sin\ 2Q$$
$$+0\overset{\circ}{.}001\ 261\ \cos\ 2\zeta\ \sin\ 2Q$$
$$+0\overset{\circ}{.}001\ 236\ \sin\ 2\zeta\ \cos\ 2Q$$
$$-0\overset{\circ}{.}002\ 075\ \cos\ 2\zeta\ \cos\ 2Q$$

URANUS

Calculate T by means of formula (23.1), and then υ, P, Q, S, W by means of the formulae given on page 110. Then calculate

$$G = 83°.76922 + 218°.4901\ T$$
$$H = 2G - S$$

The angle H varies slowly with time, increasing by 360 degrees in a period of 4229 years.

The mean elements of Uranus can be calculated by means of the coefficients given on page 100, and the planet's mean anomaly can be found from

$$M_7 = 72°.64878 + 428°.37911\ T + 0°.000\ 079\ T^2$$

Then calculate

$$\zeta = S - P, \qquad \eta = S - Q, \qquad \theta = G - S$$

$$\begin{aligned} A = \quad & (0°.864\ 319 - 0°.001\ 583\ \upsilon)\ \sin\ H \\ & + (0.082\ 222 - 0.006\ 833\ \upsilon)\ \cos\ H \\ & + 0.036\ 017\ \sin\ 2H \\ & - 0.003\ 019\ \cos\ 2H \\ & + 0.008\ 122\ \sin\ W \end{aligned}$$

$$\begin{aligned} B = \quad & 0°.120\ 303\ \sin\ H \\ & + (0.019\ 472 - 0.000\ 947\ \upsilon)\ \cos\ H \\ & + 0.006\ 197\ \sin\ 2H \end{aligned}$$

As in the case of Jupiter and Saturn, the correction to Uranus' mean longitude is A, and the correction to the mean anomaly M_7 is

$$A - \frac{B}{e}$$

e being the *un*corrected eccentricity.

Correction to the orbital eccentricity, in units of the seventh decimal :

$$+ (-3349 + 163\,\upsilon)\ \sin H$$
$$+ 20981\ \cos H$$
$$+\ \ 1311\ \cos 2H$$

Perturbation in the semimajor axis :

$$- 0.003\,825\ \cos H$$

With the mean elements thus corrected, calculate Uranus' true longitude, latitude and radius vector by means of the classical formulae (see Chapters 22 and 25). Then, add the following additive terms :

Correction to the true longitude :

$$+ (0°.010\,122 - 0°.000\,988\,\upsilon)\ \sin (S + \eta)$$
$$+ (-0.038\,581 + 0.002\,031\,\upsilon - 0.001\,910\,\upsilon^2)\ \cos (S + \eta)$$
$$+ (0.034\,964 - 0.001\,038\,\upsilon + 0.000\,868\,\upsilon^2)\ \cos (2S + \eta)$$
$$+ 0.005\,594\ \sin (S + 3\theta)$$
$$- 0.014\,808\ \sin \zeta$$
$$- 0.005\,794\ \sin \eta$$
$$+ 0.002\,347\ \cos \eta$$
$$+ 0.009\,872\ \sin \theta$$
$$+ 0.008\,803\ \sin 2\theta$$
$$- 0.004\,308\ \sin 3\theta$$

Correction to the heliocentric latitude :

$$+ (0°.000\,458\ \sin \eta - 0°.000\,642\ \cos \eta - 0°.000\,517\ \cos 4\theta)\ \sin S$$
$$- (0.000\,347\ \sin \eta + 0.000\,853\ \cos \eta + 0.000\,517\ \sin 4\eta)\ \cos S$$
$$+ 0.000\,403\ (\cos 2\theta\ \sin 2S + \sin 2\theta\ \cos 2S)$$

Correction to the radius vector, in units of the sixth decimal :

$- 25948$	$+ (5795 \cos S - 1165 \sin S + 1388 \cos 2S)\ \sin \eta$
$+ 4985 \cos \zeta$	$+ (1351 \cos S + 5702 \sin S + 1388 \sin 2S)\ \cos \eta$
$- 1230 \cos S$	$+ 904 \cos 2\theta$
$+ 3354 \cos \eta$	$+ 894\ (\cos \theta - \cos 3\theta)$

NEPTUNE

Calculate T by means of formula (23.1), then υ, P, Q, S by means of the formulae given on page 110, and then G and H as for Uranus (page 116).

The mean elements of Neptune can be calculated by means of the coefficients given on page 101, and the planet's mean anomaly can be found from

$$M_8 = 37°.73063 + 218°.46134\, T - 0°.000\,070\, T^2$$

Then calculate

$$\zeta = G - P, \qquad \eta = G - Q, \qquad \theta = G - S$$

$$
\begin{aligned}
A = \quad & (-0°.589\,833 + 0°.001\,089\ \upsilon)\ \sin H \\
+ \ & (-0.056\,094 + 0.004\,658\ \upsilon)\ \cos H \\
- \ & 0.024\,286\ \sin 2H
\end{aligned}
$$

$$
\begin{aligned}
B = & + 0°.024\,039\ \sin H \\
& - 0.025\,303\ \cos H \\
& + 0.006\,206\ \sin 2H \\
& - 0.005\,992\ \cos 2H
\end{aligned}
$$

As before, the correction to the mean longitude is A, and the correction to the mean anomaly M_8 is

$$A - \frac{B}{e}$$

e being the *un*corrected orbital eccentricity.

Correction to the orbital eccentricity, in units of the seventh decimal :

$$
\begin{array}{ll}
+\ 4389\ \sin H & +\ 1129\ \sin 2H \\
+\ 4262\ \cos H & +\ 1089\ \cos 2H
\end{array}
$$

Correction to the semimajor axis, in units of the sixth decimal :

$$- 817\ \sin H + 8189\ \cos H + 781\ \cos 2H$$

With the mean elements thus corrected, calculate Neptune's true longitude, latitude and radius vector by means of the classical formulae (see Chapters 22 and 25). Then add the following additive corrections :

Correction to the true longitude :

- $0°.009\,556$ sin ζ
- $0.005\,178$ sin η
+ $0.002\,572$ sin 2θ
- $0.002\,972$ cos 2θ sin G
- $0.002\,833$ sin 2θ cos G

Correction to the heliocentric latitude :

+ $0°.000\,336$ cos 2θ sin G
+ $0\,.000\,364$ sin 2θ cos G

Correction to the radius vector, in units of the sixth decimal :

- 40596
+ 4992 cos ζ
+ 2744 cos η
+ 2044 cos θ
+ 1051 cos 2θ

NOTES

- The periodic perturbations for the giant planets are due to Gaillot :
  JUPITER : *Annales de l'Observatoire de Paris, Mémoires* ;
  tome XXXI (1913);   SATURN : *ibid.*, tome XXIV (1904);
  URANUS and NEPTUNE : *ibid.*, tome XXVIII (1910).

- For Jupiter and Saturn, many small periodic terms have not been given here. For Uranus and for Neptune, only the most important periodic terms are given, and it is expected that for these two planets the possible error in the obtained heliocentric longitude will be of the order of 0.01 degree.

- For Pluto and the minor planets, no mean elements and periodic terms can be given, because no general "planetary theory" has been constructed for these bodies. Their future positions are calculated by numerical integration, starting from so-called osculating elements, which are valid only for a very short period.

25

Elliptic Motion

In this Chapter we will describe two methods for the calculation
of a geocentric ephemeris in the case of an elliptic orbit. In the
first method, which may be used for the major planets, the geocentric ecliptical longitude and latitude are obtained from the heliocentric ecliptical coordinates of the planet and the geocentric
longitude and radius vector of the Sun. In the second method,
which is better suited for minor planets and periodic comets, the
right ascension and declination of the body, referred to a standard equinox, are obtained directly ; use is made of the geocentric
rectangular coordinates of the Sun.

FIRST METHOD

In this method, we use the orbital elements of the planet referred
to the *mean equinox of the date.*

From Table 23.A, calculate for the given instant the planet's
mean longitude L, semimajor axis a, orbital eccentricity e, inclination i, and longitude of the ascending node Ω.

Calculate the planet's mean anomaly M by means of one of the
following formulae :

MERCURY $\quad M_1 = 102°.27\,938 + 149\,472°.51\,529\,T + 0°.000\,007\,T^2$

VENUS $\quad M_2 = 212°.603\,22 + 58\,517°.80\,387\,T + 0°.001\,286\,T^2$

MARS $\quad M_4 = 319°.51\,913 + 19\,139°.85\,475\,T + 0°.000\,181\,T^2$

JUPITER $\quad M_5 = 225°.32\,833 + 3034°.69\,202\,T - 0°.000\,722\,T^2$

SATURN $\quad M_6 = 175°.46\,622 + 1221°.55\,147\,T - 0°.000\,502\,T^2$

where T is the time in Julian centuries from 1900 January 0.5 ET ;

see formula (23.1). The cases of Uranus and Neptune are not considered here by reason of the large perturbations in their motion.

From the values of e and M, calculate the eccentric anomaly E (Chapter 22), and then the true anomaly v from

$$\tan \frac{v}{2} = \sqrt{\frac{1 + e}{1 - e}} \tan \frac{E}{2} \tag{25.1}$$

If necessary, take into account the principal perturbations (Chapter 24).

The radius vector of the planet can be calculated by means of one of the following two formulae :

$$r = a \ (1 - e \cos E)$$
$$r = \frac{a \ (1 - e^2)}{1 + e \cos v} \tag{25.2}$$

The planet's argument of latitude is

$$u = L + v - M - \Omega \tag{25.3}$$

The ecliptical longitude l can be deduced from $(l - \Omega)$, which is given by

$$\tan \ (l - \Omega) = \cos i \tan u \tag{25.4}$$

If $i < 90°$, as for the major planets, $(l - \Omega)$ and u must lie in the same quadrant. When a programmable calculator is used, in order to avoid the use of tests, formula (25.4) can better be written as follows :

$$\tan \ (l - \Omega) = \frac{\cos i \sin u}{\cos u} \tag{25.5}$$

and then the conversion from rectangular to polar coordinates should be applied to the numerator and the denominator of the fraction in the right-hand side. This will give $(l - \Omega)$ directly in the correct quadrant.

The planet's ecliptical latitude b is given by

$$\sin b = \sin u \sin i \tag{25.6}$$

with $-90° < b < +90°$.

We have now obtained the heliocentric ecliptical coordinates

l, b, r of the planet for the given instant. Its geocentric coordinates can be obtained as follows.

Using the method described in Chapter 18, calculate for the given instant the Sun's geometric longitude Θ referred to the mean equinox of the date, and its radius vector R. The planet's geocentric longitude λ can be deduced from $(\lambda - \Theta)$, which is given by

$$\tan (\lambda - \Theta) = \frac{r \cos b \sin (l - \Theta)}{r \cos b \cos (l - \Theta) + R} = \frac{N}{D} \tag{25.7}$$

Once again, $(\lambda - \Theta)$ can be obtained immediately in the correct quadrant by applying the conversion from rectangular into polar coordinates to the numerator N and the denominator D of the fraction.

The planet's distance Δ to the Earth, in astronomical units, is given by

$$\left.\begin{aligned} \Delta^2 &= N^2 + D^2 + (r \sin b)^2 \\ \text{or} \\ \Delta^2 &= R^2 + r^2 + 2rR \cos b \cos (l - \Theta) \end{aligned}\right\} \tag{25.8}$$

Finally, the planet's geocentric latitude β is given by

$$\sin \beta = \frac{r}{\Delta} \sin b \tag{25.9}$$

The geocentric coordinates of the planet obtained in this manner are the planet's *geometric* coordinates referred to the mean equinox of the date. If high accuracy is needed, it is necessary to take into account the *effect of light-time* : at time t, the planet is seen in the direction obtained by combining the Earth's (Sun's) position at time t with that of the planet at time $t - \tau$, where τ is the time taken by the light to reach the Earth from the planet. This time is given by

$$\tau = 0.005\ 7756\ \Delta \quad \text{day} \tag{25.10}$$

The *elongation* ψ of the planet, that is its angular distance to the Sun, can be calculated from

$$\cos \psi = \cos \beta \cos (\lambda - \Theta) \tag{25.11}$$

The longitude and latitude of the planet can be converted to right ascension and declination by means of the formulae (8.3) and

(8.4). The equatorial coordinates obtained in this manner are still referred to the *mean* equinox of the date. They may be converted into *apparent* right ascension and declination by correcting for nutation and aberration (see Chapter 16).

Example 25.a : Calculate the heliocentric and geocentric positions of Mercury for 1978 November 12.0 ET.

We obtain successively :

$$JD = 2443\ 824.5 \qquad\qquad E = 248°932\ 38$$

$$T = +0.788\ 624\ 230 \qquad\quad v = 238°250\ 67$$

$$L = 337°053\ 200 \qquad\qquad r = 0.415\ 71$$

$$a = 0.387\ 0986 \qquad\qquad u = 267°296\ 53$$

$$e = 0.205\ 630\ 33$$

$$i = 7°004\ 337 \qquad\qquad l - \Omega = 267°276\ 24$$

$$\Omega = 48°080\ 736 \qquad\qquad l = 315°35697 = 315°21'25''$$

$$M_1 = 259°926\ 60 \qquad\qquad b = -6°99650 = -6°59'47''$$

In Example 18.a we have found, for the same instant,

$$\Theta = 229°25049 \qquad\qquad R = 0.98984$$

Hence,

$$l - \Theta = 86°10648$$

$$\tan(\lambda - \Theta) = \frac{+0.411\ 6621}{+1.017\ 8575}$$

$$\lambda - \Theta = 22°02037$$

$$\lambda = 251°27086$$

$$\Delta = 1.09912$$

$$\beta = -2°64058$$

$$\psi = 22°17$$

By means of formula (18.4), we find $\varepsilon = 23°442\ 032$. Hence, by means of formulae (8.3) and (8.4),

$$\alpha = 249°31740 = 16^h37^m16^s2$$

$$\delta = -24°74770 = -24°44'52''$$

Let us now compare our results with the values given by the *A.E.* :

124

		Our result	A.E.
l	heliocentric longitude	315°21'25"	315°21'17"
b	heliocentric latitude	$-6°59'47"$	$-6°59'47"$
r	radius vector	0.41571	0.41572
α	right ascension	$16^h37^m16\overset{s}{.}2$	$16^h37^m14\overset{s}{.}4$
δ	declination	$-24°44'52"$	$-24°44'39"$
Δ	distance to Earth	1.09912	1.09914

The error in l is due to the fact that we neglected the perturbations in the motions of Mercury and the Earth. The errors in α
and δ are due partly to this same reason, and partly because we
neglected the effects of light-time, nutation and aberration.

SECOND METHOD

Here we use the orbital elements referred to a standard equinox,
for instance 1950.0, and the geocentric rectangular equatorial
coordinates X, Y, Z of the Sun referred to that *same* equinox. These
rectangular coordinates can be taken from the *A.E.*, or calculated
by means of the method described in Chapter 19.

The heliocentric longitude and latitude of the planet or comet
are not calculated in this method. Instead, we calculate the heliocentric rectangular equatorial coordinates x, y, z of the body,
after which the right ascension, declination and other quantities
are derived by means of simple formulae.

The following orbital elements are given :

a = semimajor axis, in AU
e = eccentricity
i = inclination
ω = argument of perihelion
Ω = longitude of ascending node
n = mean motion, in degrees/day

where i, ω and Ω are referred to a standard equinox.

If a and n are not given, they can be calculated from

$$a = \frac{q}{1 - e} \qquad\qquad n = \frac{0.985\ 609}{a \sqrt{a}} \qquad\qquad (25.12)$$

where q is the perihelion distance in AU.

All these elements are, strictly speaking, only for one given instant, called the *Epoch*. They vary slowly with time under influence of planetary perturbations. Unless high accuracy is required, the elements may be considered as invariable during several months, for example during the whole apparition of a comet.

Besides the above-mentioned orbital elements, either the value M_0 of the mean anomaly at the epoch, or the time T of passage at the perihelion is given. This allows the calculation of the mean anomaly M at any given instant. The mean anomaly increases by n degrees per day, and is zero at time T.

The orbital elements of a minor planet or of a periodic comet being given, the geocentric position for a given date can be calculated as follows. Firstly, we must calculate the quantities a, b, c and the angles A, B, C, which are constants for a given orbit.

Let ε be the obliquity of the ecliptic. If the orbital elements are referred to the standard equinox of 1950.0, one should use the value

$$\varepsilon_{1950} = 23°445\ 7889$$

Then calculate

$F = \cos \Omega$	$P = -\sin \Omega \cos i$
$G = \sin \Omega \cos \varepsilon$	$Q = \cos \Omega \cos i \cos \varepsilon - \sin i \sin \varepsilon$
$H = \sin \Omega \sin \varepsilon$	$R = \cos \Omega \cos i \sin \varepsilon + \sin i \cos \varepsilon$

As a check, we have $F^2 + G^2 + H^2 = 1$, $P^2 + Q^2 + R^2 = 1$, but this calculation is not needed in a program.

Then the quantities a, b, c, A, B, C are given by

$$\left.\begin{array}{ll} \tan A = \dfrac{F}{P} & a = \sqrt{F^2 + P^2} \\[2mm] \tan B = \dfrac{G}{Q} & b = \sqrt{G^2 + Q^2} \\[2mm] \tan C = \dfrac{H}{R} & c = \sqrt{H^2 + R^2} \end{array}\right\} \qquad (25.13)$$

The quantities a, b, c should be taken *positive*, while the angles A, B, C should be placed in the correct quadrant, according to the following rules :

sin A has the same sign as cos Ω ,

sin B and sin C have the same sign as sin Ω.

However, once again it is preferable to apply the conversion from rectangular to polar coordinates to F and P, to G and Q, and to H and R. Not only will this procedure place the angles A, B, C in the correct quadrant, but at the same time it will provide the values of a, b, c, and thus save many program steps.

For each required position, calculate the body's mean anomaly M, then the eccentric anomaly E (see Chapter 22), the true anomaly v by means of formula (25.1), and the radius vector r by means of (25.2). Then the heliocentric rectangular equatorial coordinates of the body are given by

$$\left. \begin{array}{l} x = r\,a\,\sin\,(A + \omega + v) \\[4pt] y = r\,b\,\sin\,(B + \omega + v) \\[4pt] z = r\,c\,\sin\,(C + \omega + v) \end{array} \right\} \qquad (25.14)$$

The convenience of these formulae is seen when the rectangular coordinates are required for several positions of the body. The auxiliary quantities a, b, c, A, B, C are functions only of Ω, i and ϵ, and thus are constants for the whole ephemeris ; for each position only the values of v and r must be calculated. However, it should be noted that Ω, i and ω are constant only if the body is in an unperturbed orbit.

For the same instant, calculate the Sun's rectangular coordinates X, Y, Z (Chapter 19), or take them from the *A.E.*

The geocentric right ascension α and declination δ of the planet or comet are then calculated from

$$\left. \begin{array}{l} \tan\,\alpha = \dfrac{Y + y}{X + x} \\[10pt] \Delta^2 = (X + x)^2 + (Y + y)^2 + (Z + z)^2 \\[10pt] \sin\,\delta = \dfrac{Z + z}{\Delta} \end{array} \right\} \qquad (25.15)$$

where Δ is the distance to the Earth and thus is positive. The correct quadrant of α is indicated by the fact that $\sin \alpha$ has the same sign as $(Y + y)$; however, once more, the transformation from rectangular to polar coordinates, applied to the numerator and the denominator of the fraction, will put α in the correct quadrant without any test.

If α is negative, add 360 degrees. Then transform α from degrees into hours by dividing by 15.

The elongation ψ to the Sun, and the phase angle β (the angle Sun - body - Earth), can be calculated from

$$\cos \psi = \frac{(X + x)X + (Y + y)Y + (Z + z)Z}{R\Delta} = \frac{R^2 + \Delta^2 - r^2}{2R\Delta}$$

$$\cos \beta = \frac{(X + x)x + (Y + y)y + (Z + z)z}{r\Delta} = \frac{r^2 + \Delta^2 - R^2}{2r\Delta}$$

where $R = \sqrt{X^2 + Y^2 + Z^2}$; the angles ψ and β are both between 0 and +180 degrees.

The magnitude is then calculated as follows. In the case of a *comet*, the *total* magnitude is given by

$$m = g + 5 \log \Delta + \kappa \log r \tag{25.16}$$

where g is the absolute magnitude, and κ a constant which differs from one comet to another. In general, κ is a number between 5 and 15.

In the case of a *minor planet*, we have

$$m = g + 5 \log r\Delta + k\beta$$

where β is the phase angle in degrees, and k is the phase coefficient. Generally, the value $k = 0.023$ is used for minor planets, although for some objects larger values have been found, for instance 0.049 for Ceres.

Example 25.b : Calculate the geocentric position of 433 Eros for 1975 February 11.0 ET, using the following orbital elements (IAUC 2722) :

$$
\begin{aligned}
\text{Epoch} &= 1975 \text{ January } 28.0 \text{ ET} \\
T &= 1975 \text{ January } 24.70450 \text{ ET} \\
a &= 1.457\ 9641 \text{ AU} \\
e &= 0.222\ 7021 \\
i &= 10°82772 \\
\omega &= 178°44991 \\
\Omega &= 303°83085 \\
n &= 0.559\ 865\ 65 \text{ degree/day} \\
g &= 12.4 \text{ (photographic)}
\end{aligned}
\left.\begin{array}{}\\ \\ \end{array}\right\} \text{ ecliptic and equinox } 1950.0
$$

We first calculate the auxiliary constants of the orbit :

$$
\begin{array}{ll}
F = +0.556\ 742\ 97 & P = +0.815\ 895\ 71 \\
G = -0.762\ 100\ 94 & Q = +0.426\ 938\ 36 \\
H = -0.330\ 513\ 88 & R = +0.389\ 920\ 29
\end{array}
$$

whence, by the formulae (25.13),

$$
\begin{array}{ll}
A = +34°30847 & a = 0.987\ 749\ 23 \\
B = -60°74191 & b = 0.873\ 541\ 19 \\
C = -40°28610 & c = 0.511\ 152\ 87
\end{array}
$$

For the given date (1975 February 11.0), the time from perihelion is +17.29550 days. Thus the mean anomaly is

$$
M = 17.29550 \times 0°559\ 865\ 65 = +9°683\ 156
$$

We then find

$$
\begin{array}{ll}
E = 12°429\ 591 & x = -0.841\ 5580 \\
v = 15°554\ 375 & y = +0.725\ 7529 \\
r = 1.140\ 8828 & z = +0.258\ 2179
\end{array}
$$

The Sun's geocentric rectangular equatorial coordinates for the date, referred to the same standard equinox (1950.0), are taken from the *Astronomical Ephemeris* :

$$
X = +0.770\ 0006 \qquad Y = -0.566\ 4014 \qquad Z = -0.245\ 6064
$$

We then obtain further :

$$
\begin{aligned}
X + x &= -0.071\ 5574 \\
Y + y &= +0.159\ 3515 \\
Z + z &= +0.012\ 6115 \\
R &= 0.986\ 9316 \\
\Delta &= 0.175\ 1354
\end{aligned}
$$

$$\alpha_{1950} = 114°182\ 647 = 7^h36^m44^s$$
$$\delta_{1950} = +4°07!8$$
$$\psi = 149°19$$
$$\beta = 26°30$$
$$\text{magnitude} = 9.5$$

As an exercise, calculate an ephemeris for the minor planet 234 Barbara, using the following orbital elements :

$$\text{Epoch} = 1979 \text{ November } 23.0 \text{ ET}$$
$$M_o = 34°88670$$
$$a = 2.384\ 8264$$
$$e = 0.245\ 6180$$
$$i = 15°38354$$
$$\omega = 191°11341 \left.\right\} \text{ ecliptic and}$$
$$\Omega = 144°17952 \left.\right\} \text{ equinox 1950.0}$$
$$n = 0°267\ 620\ 22$$

Compare your results with the following ephemeris, published in the *Ephemerides of Minor Planets for 1979* (Leningrad, 1978) :

0^h ET		α_{1950}	δ_{1950}
1979	Sept. 4	$1^h24^m.8$	$-\ 9°19'$
	14	1 24.6	$-12\ 14$
	24	1 21.0	$-15\ 04$
Oct.	4	1 15.2	$-17\ 30$
	14	1 08.4	$-19\ 15$
	24	1 02.2	$-20\ 11$
Nov.	3	0 57.9	$-20\ 17$
	13	0 56.2	$-19\ 39$

The Equation of the Center

If the orbital eccentricity is small, then instead of solving the equation of Kepler (Chapter 22) and then using formula (25.1), the equation of the center C, or $v - M$, can be found directly in terms of e and M by means of the following formula.

$$C = \left(2e - \frac{e^3}{4} + \frac{5}{96}e^5\right) \sin M + \left(\frac{5}{4}e^2 - \frac{11}{24}e^4\right) \sin 2M$$

$$+ \left(\frac{13}{12}e^3 - \frac{43}{64}e^5\right) \sin 3M + \frac{103}{96}e^4 \sin 4M + \frac{1097}{960}e^5 \sin 5M$$

The result is expressed in *radians*, and thus should be multiplied by $180/\pi$ or 57.295 779 51 in order to convert it in degrees. The formula is derived from a series expansion, and has been truncated after the term in e^5. Therefore it is suitable only for small values of the eccentricity. If the eccentricity is *very* small, the terms in e^4 and e^5 may be neglected.

The greatest error is

	The formula up to terms in e^5	The formula with terms e^4 and e^5 neglected
for $e = 0.03$	0".0003	0".24
0.05	0.007	1.8
0.10	0.45	30
0.15	5	152
0.20	29	483
0.25	111	1183
0.30	331	2456

Example 25.c : Take, as in Example 25.a, $e = 0.205\,630\,33$ and $M = 259°.926\,60$.

The formula above then gives $C = -0.378\,459\,88$ radian
$$= -21°.684\,15$$

whence $v = M + C = 238°.242\,45$.

The correct value, found in Example 25.a, is $238°.250\,67$. Hence, in this case the error is 0.00822 degree or 30".

26

PARABOLIC MOTION

In this Chapter we will give formulae for the calculation of positions of a comet which moves around the Sun in a parabolic orbit. We will assume that the elements of this orbit are invariable (no planetary perturbations) and that they are referred to a standard equinox (for example 1950.0).

The following orbital elements are given :

- T = time of passage in perihelion
- q = perihelion distance, in AU
- i = inclination
- ω = argument of perihelion
- Ω = longitude of ascending node

Firstly, calculate the auxiliary constants a, b, c, A, B, C as for an elliptic orbit : see formulae (25.13). Then, for each required position of the comet, proceed as follows.

Let $t - T$ be the time since perihelion, in days. This quantity is negative for an instant before the time of perihelion. Calculate

$$W = \frac{0.036\ 491\ 1624}{q\ \sqrt{q}}\ (t - T) \tag{26.1}$$

Then the true anomaly v and the radius vector r of the comet are given by

$$\tan \frac{v}{2} = s \qquad\qquad r = q\ (1 + s^2) \tag{26.2}$$

where s is the root of the equation

$$s^3 + 3s - W = 0 \tag{26.3}$$

This equation can easily be solved by iteration. One may start from *any* value ; a good choice is $s = 0$. A better value for s is then given by

$$\frac{2s^3 + W}{3(s^2 + 1)} \tag{26.4}$$

This calculation is repeated until the correct value of s is obtained. It should be noted that in formula (26.4) the cube of s must be calculated ; if s is negative, this operation is not possible on some calculating machines ; when this is the case, calculate $s^2 \times s$ instead of s^3.

Instead of solving equation (26.3) by iteration, s can be obtained directly as follows (J. Bauschinger, *Tafeln zur Theoretischen Astronomie*, page 9 ; Leipzig, 1934) :

$$\tan \beta = \frac{2}{W} = 54.807\ 791\ \frac{q\sqrt{q}}{t - T}$$

$$\tan \gamma = \sqrt[3]{\tan \frac{\beta}{2}} \tag{26.5}$$

$$s = \frac{2}{\tan 2\gamma}$$

For the calculation of $\tan \gamma$, one must take the cube root of a quantity which may be negative. When this is the case, the operation is impossible on most calculating machines. This difficulty can be avoided by using a test, a flag, or any other trick. For instance, here are two procedures for calculating the cube root of *any* number on the HP-67 calculating machine :

First method	*Second method*
f $x < 0$	h ABS
h SF 2	h LST x
h ABS	:
3	h LST x
h $1/x$	h ABS
h y^x	3
h F? 2	h $1/x$
CHS	h y^x
	×

However, the author's preference is the iteration formula (26.4) which works without any difficulty.

When s is obtained, v and r can be found by means of (26.2), after which the calculation continues as for the elliptic motion : formulae (25.14) and (25.15).

It should be noted that s has the same sign as $t - T$, and thus is negative before perihelion, positive after perihelion.

In the parabolic motion, $e = 1$ while a and the period of revolution are infinite ; the mean daily motion is zero, and therefore the mean and eccentric anomalies do not exist (in fact, they are zero).

Example 26.a : Calculate the geocentric position of comet Kohler (1977m) for 1977 September 29.0 ET, using the following parabolic elements (IAUC 3137) :

$$T = \text{1977 November 10.5659 ET}$$
$$q = 0.990\,662$$
$$\left.\begin{array}{l} i = 48°7196 \\ \omega = 163.4799 \\ \Omega = 181.8175 \end{array}\right\} 1950.0$$

magnitude = $6.0 + 5 \log \Delta + 10 \log r$

We first calculate the auxiliary constants of the orbit :

$F = -0.999\,496\,92$	$P = +0.020\,924\,49$
$G = -0.029\,097\,47$	$Q = -0.903\,973\,29$
$H = -0.012\,619\,22$	$R = +0.427\,076\,64$

whence, by the formulae (25.13),

$A = -88°800\,69$	$a = 0.999\,715\,92$
$B = -178°156\,38$	$b = 0.904\,441\,47$
$C = -1°692\,48$	$c = 0.427\,263\,04$

For the given date (1977 September 29.0), the time from perihelion is $t - T = -42.5659$ days. Hence, by formula (26.1),

$$W = -1.575\,2927$$

Starting from the value $s = 0$, we obtain the following successive approximations by means of the iteration formula (26.4) :

$$0.000\,0000$$
$$-0.525\,0976$$
$$-0.487\,2672$$
$$-0.486\,6745$$
$$-0.486\,6743$$

Hence, $s = -0.486\,6743$, and consequently

$$v = -51°90199 \qquad r = 1.225\ 3022$$

If, instead of the iteration procedure, formulae (26.5) are preferred, we obtain successively :

$$\tan \beta = -1.269\ 6053$$
$$\beta = -51°774\ 3927$$

$$\tan \gamma = \sqrt[3]{-0.485\ 2978} = -0.785\ 8436$$
$$\gamma = -38°161\ 8063$$
$$s = -0.486\ 6743, \text{ as before.}$$

We then find, by means of formulae (25.14),

$$x = +0.474\ 2398$$
$$y = -1.016\ 9032$$
$$z = +0.492\ 3109$$

The Sun's geocentric rectangular equatorial coordinates for the given date, referred to the same standard equinox (1950.0), are taken from the *Astronomical Ephemeris* :

$$X = -0.997\ 3057$$
$$Y = -0.085\ 7667$$
$$Z = -0.037\ 1837$$

We then obtain further :

$$X + x = -0.523\ 0659$$
$$Y + y = -1.102\ 6699 \qquad\qquad R = 1.001\ 6772$$
$$Z + z = +0.455\ 1272 \qquad\qquad \Delta = 1.302\ 5435$$

$$\alpha_{1950} = -115°377\ 936 = 16^h 18^m 29^s$$
$$\delta_{1950} = +20°27'.1$$

$$\psi = 62°66$$
$$\text{magnitude} = 7.5$$

27

PLANETS IN PERIHELION AND APHELION

The Julian Day corresponding to the time when a planet is in perihelion or aphelion can be found by means of the following formulae :

Mercury	JD = 2414 995.007 + 87.969 349 97 k
Venus	JD = 2415 112.001 + 224.700 8454 k − 0.000 000 0304 k^2
Earth	JD = 2415 021.546 + 365.259 6413 k + 0.000 000 0152 k^2
Mars	JD = 2415 097.251 + 686.995 8091 k − 0.000 000 1221 k^2
Jupiter	JD = 2416 640.884 + 4332.894 375 k + 0.000 1222 k^2
Saturn	JD = 2409 773.47 + 10 764.180 10 k + 0.001 3033 k^2

where k is an integer for perihelion, and an integer increased by exactly 0.5 for aphelion.

Any other value for k will give a meaningless result !

A positive (negative) value of k will give a date after (before) the beginning of the year 1900.

For example, $k = +14$ and $k = -222$ will give passages through perihelion, while $k = +27.5$ and $k = -119.5$ give passages through aphelion.

An *approximate* value for k can be found as follows, where the "year" should be taken with decimals, if necessary :

Mercury	$k \simeq 4.15201$ (year − 1900)
Venus	$k \simeq 1.62549$ (year − 1900)
Earth	$k \simeq 0.99997$ (year − 1900)
Mars	$k \simeq 0.53166$ (year − 1900)

| Jupiter | $k \simeq 0.08430$ (year $-$ 1900) |
| Saturn | $k \simeq 0.03393$ (year $-$ 1900) |

Example 27.a : Find the time of passage of Venus at the perihe-lion nearest to 1978 October 15, that is 1978.79.

An approximate value of k is given by

$$k \simeq 1.62549 \, (1978.79 - 1900) = 128.07$$

and, since k must be an integer (perihelion !), we take $k = 128$. Putting this value in the formula for Venus, we find

$$JD = 2443\,873.709\,,$$

which corresponds to 1978 December 31.209 = 1978 December 31 at 5^h ET.

Example 27.b : Find the time of passage of Mars through aphelion in 1978.

Taking year = 1978, we find $k \simeq 41.47$. Since k must be an integer increased by 0.5 (aphelion !), we take $k = 41.5$.
Using the formula for Mars, this gives $JD = 2443\,607.577$, which corresponds to 1978 April 9.077 or 1978 April 9 at 2^h ET.

It is important to note that the formula given for the Earth is actually valid for the *barycenter* of the Earth-Moon system. Due to the action of the Moon, the time of least or greatest dis-tance between the centers of Sun and Earth may differ from that for the barycenter by more than one day. For instance, $k = 78$ in the formula for the Earth yields JD = 2443 511.80, which corres-ponds to 1978 January 3.30, while the correct instant for the Earth is January 1 at 23^h.

Due to mutual planetary perturbations, the instants for Jupiter, calculated by the method described here, may be up to half a month in error. For Saturn, the error may be larger than one month.

For instance, putting $k = 6.5$ in the formula for Jupiter gives 1981 July 19 as the date of an aphelion passage, while the correct date is 1981 July 28. For Saturn, $k = 2$ gives 1944 July 30, while the planet actually reached perihelion on 1944 September 8.

The error would be even larger for Uranus and Neptune. For this reason, no formula is given for these planets. Uranus reached the perihelion on 1966 May 19, and will be in aphelion on 2009 Feb 27.

28

Passages through the Nodes

Given the orbital elements of a planet or comet, the times t of passages of that body through the nodes of its orbit can easily be calculated as follows.

We have

at the ascending node : $v = -\omega$ or $360° - \omega$

at the descending node : $v = 180° - \omega$

where, as before, v is the true anomaly, and ω the argument of the perihelion. Then, with these values of v, proceed as follows :

Case of an elliptic orbit

$$\tan \frac{E}{2} = \sqrt{\frac{1 - e}{1 + e}} \tan \frac{v}{2}$$

$$M = E - e_o \sin E \tag{28.1}$$

$$t = T + \frac{M}{n} \text{ days} \tag{28.2}$$

where e is the orbital eccentricity, while e_o is e converted from radians into degrees, that is

$$e_o = e \times 57°.295\ 779\ 51$$

In formula (28.1), E should be expressed in degrees. In formula (28.2), T is the time of perihelion passage, M is expressed in degrees, while n is the mean motion in degrees/day.

The corresponding value of the radius vector r can be calculated from

$$r = a\ (1 - e \cos E)$$

where a is the semimajor axis, expressed in astronomical units.

If a and n are not given, they can be calculated from

$$a = \frac{q}{1 - e} \qquad\qquad n = \frac{0.985\,609}{a\sqrt{a}}$$

where q is the perihelion distance in astronomical units.

Case of a parabolic orbit

$$s = \tan\frac{v}{2}$$

$$t = T + 27.403\,896\,(s^3 + 3s)\,q\sqrt{q}\ \text{ days}$$

where the perihelion distance q is expressed in AU. The corresponding value of the radius vector is

$$r = q\,(1 + s^2)$$

Note. — The nodes refer to the ecliptic of the same epoch as that of the equinox used for the orbital elements. For example, if the orbital elements are referred to the standard equinox of 1950.0, the above-mentioned formulae give the passages through the nodes on the ecliptic of 1950.0, *not* on the ecliptic of date. The difference may generally be neglected, except when the inclination is very small.

Example 28.a :   We use the same orbital elements for the minor planet Eros as in Example 25.b :

$$T = 1975 \text{ January } 24.70450 \text{ ET}$$
$$\omega = 178°\!.44991$$
$$e = 0.222\,7021$$
$$n = 0°\!.559\,865\,65 \text{ per day}$$
$$a = 1.457\,9641 \text{ AU}$$

For the passage at the descending node, we have

$$v = 180° - \omega = 1°\!.55009$$

$$\tan\frac{E}{2} = 0.797\,3214 \times 0.013\,5279 = 0.010\,7861$$

$$E = 1°\!.235\,9474$$

$$M = 1°\!.235\,9474 - (0.222\,7021 \times 57°\!.295\,779\,51)\,\sin 1°\!.2359474$$
$$= 0°\!.960\,7206$$

$$t = T + \frac{0.960\,7206}{0.559\,865\,65} = T + 1.71598 \text{ days}$$
$$= 1975 \text{ January } 26.4205$$

$$r = 1.13335 \text{ AU}$$

For the ascending node we find similarly :

$$v = -\omega = -178°44991$$
$$E = -178°05595$$
$$M = -177°62308$$
$$t = T - 317.26019 \text{ days} = 1974 \text{ March } 13.4443$$
$$r = 1.78247 \text{ AU}$$

Example 28.b : Comet Kohler (1977m). We use the same orbital elements as in Example 26.a :

$$T = 1977 \text{ November } 10.5659 \text{ ET}$$
$$q = 0.990\ 662 \text{ AU}$$
$$\omega = 163°4799$$

For the ascending node, we have

$$v = -\omega = -163°4799$$
$$s = -6.888\ 371$$
$$t = T - 9390.2 \text{ days}$$
$$= 1952 \text{ February } 25$$
$$r = 47.997 \text{ AU}$$

At the descending node we have

$$v = 180° - \omega = 16°5201$$
$$s = +0.145\ 1722$$
$$t = T + 11.8507 \text{ days}$$
$$= 1977 \text{ November } 22.4166 \text{ ET}$$
$$r = 1.0115 \text{ AU}$$

Example 28.c : Calculate the time of passage of Venus at the ascending node nearest to the epoch 1979.0.

We use the elements given in Table 23.A. There we find

$$a = 0.723\ 3316, \quad \text{whence} \quad n = 1.602\ 133$$
$$e = 0.006\ 820\ 69 - 0.000\ 047\ 74\ T + 0.000\ 000\ 091\ T^2$$
$$\omega = 54°384\ 186 + 0°508\ 1861\ T - 0.001\ 3864\ T^2$$

The elements e and ω vary with time. We calculate their values for the epoch 1979.0, that is for $T = +0.79$. We find

$$e = 0.006\ 783\ 03 \qquad \omega = 54°784\ 788$$

We then find successively

$$v = -\omega = -54°784\ 788 \qquad M = -54°151\ 620$$
$$E = -54°467\ 890 \qquad t = T - 33.7997 \text{ days}$$

In Example 27.a, we have found $T = 1978$ December 31.209 for the time of passage of Venus in the perihelion. Therefore, we have

$$t = 1978 \text{ November } 27.409 \quad \text{or} \quad 1978 \text{ November } 27 \text{ at } 10^h \text{ ET.}$$

29

Correction for Parallax

We wish to calculate the topocentric coordinates of a body (Moon, Sun, planet, comet) when its geocentric coordinates are known. *Geocentric* = as seen from the center of the Earth ; *topocentric* = as seen from the observer's place (Greek : *topos* = place ; compare with Topology).

In other words, we wish to find the corrections $\Delta\alpha$ and $\Delta\delta$ (the parallaxes in right ascension and declination), in order to obtain the topocentric right ascension $\alpha' = \alpha + \Delta\alpha$ and the topocentric declination $\delta' = \delta + \Delta\delta$, when the geocentric values α and δ are known.

Let ρ be the geocentric radius and ϕ' the geocentric latitude of the observer. The expressions $\rho \sin \phi'$ and $\rho \cos \phi'$ can be calculated by the method described in Chapter 6.

Let π be the equatorial horizontal parallax of the body. For the Sun, planets and comets, it is frequently more convenient to use the distance Δ (in astronomical units) to the Earth instead of the parallax. We then have

$$\sin \pi = \frac{\sin 8\rlap{.}''794}{\Delta}$$

or, with sufficient accuracy,

$$\pi = \frac{8\rlap{.}''794}{\Delta} \qquad (29.1)$$

Then, if H is the geocentric hour angle, the rigorous formulae are :

$$\tan \Delta\alpha = \frac{-\rho \cos \phi' \sin \pi \sin H}{\cos \delta - \rho \cos \phi' \sin \pi \cos H} \qquad (29.2)$$

In the case of the declination we may, instead of computing $\Delta\delta$, calculate δ' directly from

$$\tan\delta' = \frac{(\sin\delta - \rho\sin\phi'\sin\pi)\cos\Delta\alpha}{\cos\delta - \rho\cos\phi'\sin\pi\cos H} \qquad (29.3)$$

Except for the Moon, the following non-rigorous formulae may often be used instead of (29.2) and (29.3) :

$$\Delta\alpha = \frac{-\pi\,\rho\cos\phi'\sin H}{\cos\delta} \qquad (29.4)$$

$$\Delta\delta = -\pi\,(\rho\sin\phi'\cos\delta - \rho\cos\phi'\cos H\sin\delta) \qquad (29.5)$$

If π is expressed in seconds of a degree ($''$), then $\Delta\alpha$ and $\Delta\delta$ too are expressed in this unit. To express $\Delta\alpha$ in seconds of time, divide the result by 15.

Example 29.a :   Calculate the topocentric coordinates of Mars on 1971 August 12 at $2^h34^m00^s$ UT at the Uccle Observatory, for which

$$\rho\sin\phi' = +0.771\,306$$
$$\rho\cos\phi' = +0.633\,333$$
$$L = \text{longitude} = -0^h17^m26^s$$

Mars' geocentric equatorial coordinates for the given instant, interpolated from the *Astronomical Ephemeris*, are

$$\alpha = 21^h24^m46^s\!.85, \qquad \delta = -22°24'09''\!.9$$

The planet's distance at that time is 0.3757 AU and thus, by formula (29.1), its equatorial horizontal parallax is $\pi = 23''\!.41$.

We still need the geocentric hour angle, which is equal to $H = \theta_o - L - \alpha$, where θ_o, the sidereal time at Greenwich, can be found as indicated in Chapter 7. For the given instant we find $\theta_o = 23^h53^m36^s$. Thus

$$H = 23^h53^m36^s + 0^h17^m26^s - 21^h24^m47^s$$
$$= +2^h46^m15^s = +41°\!.562$$

Formula (29.2) then gives

$$\tan\Delta\alpha = \frac{-0.000\,047\,687}{+0.924\,474}$$

whence $\Delta\alpha$ = $-0\overset{s}{.}002\ 9555$ = $-0\overset{s}{.}71$

α' = $\alpha + \Delta\alpha$ = $21^h24^m46\overset{s}{.}14$

Formula (29.3) gives

$$\tan \delta' = \frac{-0.381\ 202\ 29}{+0.924\ 473\ 96}$$

whence δ' = $-22°24'30\overset{''}{.}8$

If, instead of (29.2) and (29.3), we choose the non-rigorous formulae (29.4) and (29.5), we find

$\Delta\alpha = -10\overset{''}{.}64 = -0\overset{s}{.}71$, as above ;

$\Delta\delta = -20\overset{''}{.}9$, whence $\delta' = \delta - 20\overset{''}{.}9 = -22°24'30\overset{''}{.}8$,
as above.

As an exercise, perform the calculation for the Moon, again for the Uccle Observatory, using fictive values, for example

$\alpha = 1^h00^m00\overset{s}{.}00 = 15\overset{°}{.}000\ 000$
$\delta = +5\overset{°}{.}000\ 000$
$H = +4^h00^m00\overset{s}{.}00 = +60\overset{°}{.}000\ 000$
$\pi = 0°59'00''$

Compare the results of the rigorous formulae with those of the non-rigorous ones.

We can consider the opposite problem : from the observed topocentric coordinates α' and δ' , deduce the geocentric values α and δ. In the case of a planet or comet, the corrections $\Delta\alpha$ and $\Delta\delta$ are so small, that the formulae (29.4) and (29.5) can be used also for the reduction from topocentric to geocentric coordinates.

Parallax in ecliptical coordinates

It is possible to calculate the topocentric coordinates of a celestial body, from its geocentric values, directly in ecliptical coordinates. The following formulae were given by Joseph Johann von Littrow (*Theoretische und Practische Astronomie*, Vol. I, p. 91 ; Wien, 1821).

Let λ = geocentric ecliptical longitude of the body (Moon, planet, comet),
β = its geocentric ecliptical latitude,
s = its geocentric semidiameter,
λ', β', s' = the required topocentric values of the same quantities,
ϕ = the observer's latitude,
ε = the obliquity of the ecliptic,
θ = the local sidereal time.

$$\phi' = \phi - 0°193 \sin 2\phi$$

$$N = \cos \lambda \cos \beta - \sin \pi \cos \phi' \cos \theta$$

$$\tan \lambda' = \frac{\sin \lambda \cos \beta - \sin \pi (\sin \phi' \sin \varepsilon + \cos \phi' \cos \varepsilon \sin \theta)}{N}$$

$$\tan \beta' = \frac{\cos \lambda' \left(\sin \beta - \sin \pi (\sin \phi' \cos \varepsilon - \cos \phi' \sin \varepsilon \sin \theta)\right)}{N}$$

$$\sin s' = \frac{\cos \lambda' \cos \beta' \sin s}{N}$$

As an exercise, calculate λ', β', s' from the following data :

$$\lambda = 181°46'22''5 \qquad \varepsilon = 23°28'00''8$$
$$\beta = +2°17'26''2 \qquad \theta = 209°46'07''9$$
$$\pi = 0°59'27''7 \qquad \phi = +50°05'07''8$$
$$s = 0°16'15''5$$

Answer :
$$\lambda' = 181°48'05''2$$
$$\beta' = +1°29'01''3$$
$$s' = 0°16'25''5$$

30

POSITION OF THE MOON

In order to calculate an accurate position of the Moon, it is necessary to take into account *hundreds* of periodic terms in the
Moon's longitude, latitude and parallax. For this reason, we will
limit ourselves to the most important periodic terms, and be sa-
tisfied with an accuracy of about $10''$ in the longitude of the Moon,
$3''$ in its latitude, and $0.''2$ in its parallax.

Using the method described below, one obtains the geocentric
longitude λ and the geocentric latitude β of the center of the
Moon, referred to the mean equinox of the date. If necessary,
λ and β can be converted to α and δ using formulae (8.3) and (8.4).
The equatorial horizontal parallax π of the Moon too is obtained.

When the parallax π is known, the distance between the centers
of Earth and Moon can be found from

$$D = \frac{1}{\sin \pi} \quad \text{equatorial radii of the Earth}$$

$$\text{or} \quad D = \frac{6378.14}{\sin \pi} \quad \text{kilometers}$$

For the given instant (ET !), calculate the JD (see Chapter 3),
and then T by means of formula (15.1). Remember that T is expressed in centuries, and thus should be taken with a sufficient number of decimals (at least nine, since during $0.000\,000\,001$ century
the Moon moves over an arc of $1.''7$).

Then calculate the angles L', M, M', D, F and Ω by means of
the following formulae, in which the various constants are expressed in degrees and decimals.

Moon's mean longitude :

$$L' = 270.434\,164 + 481\,267.8831\,T - 0.001\,133\,T^2 + 0.000\,0019\,T^3$$

Sun's mean anomaly :

$$M = 358.475\,833 + 35\,999.0498\,T - 0.000\,150\,T^2 - 0.000\,0033\,T^3$$

Moon's mean anomaly :

$$M' = 296.104\,608 + 477\,198.8491\,T + 0.009\,192\,T^2 + 0.000\,0144\,T^3$$

Moon's mean elongation :

$$D = 350.737\,486 + 445\,267.1142\,T - 0.001\,436\,T^2 + 0.000\,0019\,T^3$$

Mean distance of Moon from its ascending node :

$$F = 11.250\,889 + 483\,202.0251\,T - 0.003\,211\,T^2 - 0.000\,0003\,T^3$$

Longitude of Moon's ascending node :

$$\Omega = 259.183\,275 - 1934.1420\,T + 0.002\,078\,T^2 + 0.000\,0022\,T^3$$

To the mean values of these arguments must be added some periodic variations, called "additive terms" :

Additive to	*Term*
L'	$+0\overset{s}{.}000\,233 \; \sin\,(51\overset{\circ}{.}2 + 20\overset{\circ}{.}2\,T)$
M	$-0\overset{s}{.}001\,778 \; \sin\,(51\overset{\circ}{.}2 + 20\overset{\circ}{.}2\,T)$
M'	$+0\overset{s}{.}000\,817 \; \sin\,(51\overset{\circ}{.}2 + 20\overset{\circ}{.}2\,T)$
D	$+0\overset{s}{.}002\,011 \; \sin\,(51\overset{\circ}{.}2 + 20\overset{\circ}{.}2\,T)$
$L',\ M',\ D,\ F$	$+0\overset{s}{.}003\,964 \; \sin\,(346\overset{\circ}{.}560 + 132\overset{\circ}{.}870\,T - 0\overset{\circ}{.}009\,1731\,T^2)$
L'	$+0\overset{s}{.}001\,964 \; \sin\,\Omega$
M'	$+0\overset{s}{.}002\,541 \; \sin\,\Omega$
D	$+0\overset{s}{.}001\,964 \; \sin\,\Omega$
F	$-0\overset{s}{.}024\,691 \; \sin\,\Omega$
F	$-0\overset{s}{.}004\,328 \; \sin\,(\Omega + 275\overset{\circ}{.}05 - 2\overset{\circ}{.}30\,T)$

The first four terms have a period of 1782 years. The fifth term, with coefficient $0\overset{s}{.}003\,964$, is the "Great Venus Term" ; its period is 271 years.

With the values of L', M, M', D and F, corrected for the additive terms, λ, β and π can be obtained by means of the following expressions where, again, all the coefficients are given in degrees and decimals. The terms indicated by *(e)* or *(e²)* should be multiplied by e or e^2, where

$$e = 1 - 0.002\,495\,T - 0.000\,007\,52\,T^2$$

	$\lambda = L' + 6.288\,750 \sin M'$
	$+\ 1.274\,018 \sin (2D - M')$
	$+\ 0.658\,309 \sin 2D$
	$+\ 0.213\,616 \sin 2M'$
(e)	$-\ 0.185\,596 \sin M$
	$-\ 0.114\,336 \sin 2F$
	$+\ 0.058\,793 \sin (2D - 2M')$
(e)	$+\ 0.057\,212 \sin (2D - M - M')$
	$+\ 0.053\,320 \sin (2D + M')$
(e)	$+\ 0.045\,874 \sin (2D - M)$
(e)	$+\ 0.041\,024 \sin (M' - M)$
	$-\ 0.034\,718 \sin D$
(e)	$-\ 0.030\,465 \sin (M + M')$
	$+\ 0.015\,326 \sin (2D - 2F)$
	$-\ 0.012\,528 \sin (2F + M')$
	$-\ 0.010\,980 \sin (2F - M')$
	$+\ 0.010\,674 \sin (4D - M')$
	$+\ 0.010\,034 \sin 3M'$
	$+\ 0.008\,548 \sin (4D - 2M')$
(e)	$-\ 0.007\,910 \sin (M - M' + 2D)$
(e)	$-\ 0.006\,783 \sin (2D + M)$
	$+\ 0.005\,162 \sin (M' - D)$
(e)	$+\ 0.005\,000 \sin (M + D)$
(e)	$+\ 0.004\,049 \sin (M' - M + 2D)$
	$+\ 0.003\,996 \sin (2M' + 2D)$
	$+\ 0.003\,862 \sin 4D$
	$+\ 0.003\,665 \sin (2D - 3M')$
(e)	$+\ 0.002\,695 \sin (2M' - M)$
	$+\ 0.002\,602 \sin (M' - 2F - 2D)$
(e)	$+\ 0.002\,396 \sin (2D - M - 2M')$
	$-\ 0.002\,349 \sin (M' + D)$
(e²)	$+\ 0.002\,249 \sin (2D - 2M)$
(e)	$-\ 0.002\,125 \sin (2M' + M)$
(e²)	$-\ 0.002\,079 \sin 2M$
(e²)	$+\ 0.002\,059 \sin (2D - M' - 2M)$
	$-\ 0.001\,773 \sin (M' + 2D - 2F)$
	$-\ 0.001\,595 \sin (2F + 2D)$

(e) $+ 0.001\,220 \sin (4D - M - M')$
  $- 0.001\,110 \sin (2M' + 2F)$
  $+ 0.000\,892 \sin (M' - 3D)$
(e) $- 0.000\,811 \sin (M + M' + 2D)$
(e) $+ 0.000\,761 \sin (4D - M - 2M')$
(e^2) $+ 0.000\,717 \sin (M' - 2M)$
(e^2) $+ 0.000\,704 \sin (M' - 2M - 2D)$
(e) $+ 0.000\,693 \sin (M - 2M' + 2D)$
(e) $+ 0.000\,598 \sin (2D - M - 2F)$
  $+ 0.000\,550 \sin (M' + 4D)$
  $+ 0.000\,538 \sin 4M'$
(e) $+ 0.000\,521 \sin (4D - M)$
  $+ 0.000\,486 \sin (2M' - D)$

$$B = + 5.128\,189 \sin F$$
  $+ 0.280\,606 \sin (M' + F)$
  $+ 0.277\,693 \sin (M' - F)$
  $+ 0.173\,238 \sin (2D - F)$
  $+ 0.055\,413 \sin (2D + F - M')$
  $+ 0.046\,272 \sin (2D - F - M')$
  $+ 0.032\,573 \sin (2D + F)$
  $+ 0.017\,198 \sin (2M' + F)$
  $+ 0.009\,267 \sin (2D + M' - F)$
  $+ 0.008\,823 \sin (2M' - F)$
(e) $+ 0.008\,247 \sin (2D - M - F)$
  $+ 0.004\,323 \sin (2D - F - 2M')$
  $+ 0.004\,200 \sin (2D + F + M')$
(e) $+ 0.003\,372 \sin (F - M - 2D)$
(e) $+ 0.002\,472 \sin (2D + F - M - M')$
(e) $+ 0.002\,222 \sin (2D + F - M)$
(e) $+ 0.002\,072 \sin (2D - F - M - M')$
(e) $+ 0.001\,877 \sin (F - M + M')$
  $+ 0.001\,828 \sin (4D - F - M')$
(e) $- 0.001\,803 \sin (F + M)$
  $- 0.001\,750 \sin 3F$
(e) $+ 0.001\,570 \sin (M' - M - F)$
  $- 0.001\,487 \sin (F + D)$
(e) $- 0.001\,481 \sin (F + M + M')$
(e) $+ 0.001\,417 \sin (F - M - M')$
(e) $+ 0.001\,350 \sin (F - M)$
  $+ 0.001\,330 \sin (F - D)$
  $+ 0.001\,106 \sin (F + 3M')$
  $+ 0.001\,020 \sin (4D - F)$
  $+ 0.000\,833 \sin (F + 4D - M')$

$$+\ 0.000\,781\ \sin\ (M' - 3F)$$
$$+\ 0.000\,670\ \sin\ (F + 4D - 2M')$$
$$+\ 0.000\,606\ \sin\ (2D - 3F)$$
$$+\ 0.000\,597\ \sin\ (2D + 2M' - F)$$
$$(e)\qquad +\ 0.000\,492\ \sin\ (2D + M' - M - F)$$
$$+\ 0.000\,450\ \sin\ (2M' - F - 2D)$$
$$+\ 0.000\,439\ \sin\ (3M' - F)$$
$$+\ 0.000\,423\ \sin\ (F + 2D + 2M')$$
$$+\ 0.000\,422\ \sin\ (2D - F - 3M')$$
$$(e)\qquad -\ 0.000\,367\ \sin\ (M + F + 2D - M')$$
$$(e)\qquad -\ 0.000\,353\ \sin\ (M + F + 2D)$$
$$+\ 0.000\,331\ \sin\ (F + 4D)$$
$$(e)\qquad +\ 0.000\,317\ \sin\ (2D + F - M + M')$$
$$(e^2)\qquad +\ 0.000\,306\ \sin\ (2D - 2M - F)$$
$$-\ 0.000\,283\ \sin\ (M' + 3F)$$

$$\omega_1 = 0.000\,4664\ \cos\ \Omega$$

$$\omega_2 = 0.000\,0754\ \cos\ (\Omega + 275°\!.05 - 2°\!.30\ T)$$

$$\beta = B \times (1 - \omega_1 - \omega_2)$$

$$\pi = \mathbf{0}.950\,724$$
$$+\ 0.051\,818\ \cos\ M'$$
$$+\ 0.009\,531\ \cos\ (2D - M')$$
$$+\ 0.007\,843\ \cos\ 2D$$
$$+\ 0.002\,824\ \cos\ 2M'$$
$$+\ 0.000\,857\ \cos\ (2D + M')$$
$$(e)\qquad +\ 0.000\,533\ \cos\ (2D - M)$$
$$(e)\qquad +\ 0.000\,401\ \cos\ (2D - M - M')$$
$$(e)\qquad +\ 0.000\,320\ \cos\ (M' - M)$$
$$-\ 0.000\,271\ \cos\ D$$
$$(e)\qquad -\ 0.000\,264\ \cos\ (M + M')$$
$$-\ 0.000\,198\ \cos\ (2F - M')$$
$$+\ 0.000\,173\ \cos\ 3M'$$
$$+\ 0.000\,167\ \cos\ (4D - M')$$
$$(e)\qquad -\ 0.000\,111\ \cos\ M$$
$$+\ 0.000\,103\ \cos\ (4D - 2M')$$
$$-\ 0.000\,084\ \cos\ (2M' - 2D)$$
$$(e)\qquad -\ 0.000\,083\ \cos\ (2D + M)$$
$$+\ 0.000\,079\ \cos\ (2D + 2M')$$
$$+\ 0.000\,072\ \cos\ 4D$$

(e)	+ 0.000 064 cos (2D − M + M')
(e)	− 0.000 063 cos (2D + M − M')
(e)	+ 0.000 041 cos (M + D)
(e)	+ 0.000 035 cos (2M' − M)
	− 0.000 033 cos (3M' − 2D)
	− 0.000 030 cos (M' + D)
	− 0.000 029 cos (2F − 2D)
(e)	− 0.000 029 cos (2M' + M)
(e²)	+ 0.000 026 cos (2D − 2M)
	− 0.000 023 cos (2F − 2D + M')
(e)	+ 0.000 019 cos (4D − M − M')

Example 30.a : Calculate the geocentric longitude, latitude and equatorial horizontal parallax of the Moon on 1979 Dec. 7.0 ET.

We find successively :

$$JD = 2444\ 214.5$$
$$T = +0.799\ 301\ 8480$$

$L' = 108°7418$	$M = 332°5828$	$M' = 122°0324$
$D = 213.5638$	$F = 315.5204$	$\Omega = 153.2213$

With additive terms :

$L' = 108°7469$	$M = 332°5812$	$M' = 122°0383$
$D = 213.5705$	$F = 315.5093$	
	$e = 0.998\ 001$	

Then the Moon's longitude is equal to the sum of the following quantities :

108°7469	+ 0.020 806	− 0.005 160	+ 0.001 266	+ 0.000 660
+ 5.330 934	+ 0.019 198	− 0.000 535	+ 0.001 693	+ 0.000 285
− 1.042 303	− 0.030 305	− 0.002 409	− 0.000 002	− 0.000 037
+ 0.606 608	+ 0.006 204	− 0.003 006	+ 0.001 755	− 0.000 534
− 0.192 122	− 0.006 834	+ 0.002 765	+ 0.000 593	+ 0.000 423
+ 0.085 294	− 0.005 658	+ 0.003 206	+ 0.000 777	+ 0.000 163
+ 0.114 318	+ 0.002 264	− 0.002 689	− 0.000 467	+ 0.000 247
− 0.003 143	+ 0.001 069	+ 0.001 534	− 0.000 324	
− 0.026 346	− 0.008 043	− 0.001 213	− 0.000 253	
− 0.008 506	+ 0.007 823	+ 0.000 970	− 0.000 753	
+ 0.045 637	− 0.004 326	+ 0.001 900	+ 0.000 039	

Hence, $\lambda = 113°6604 = 113°39'37''$

In the same manner, we find :

$$B = -3\overset{''}{.}162\ 450$$
$$\omega_1 = -0.000\ 4164$$
$$\omega_2 = +0.000\ 0301$$

$$\beta = -3\overset{''}{.}162\ 450 \times 1.000\ 3863 = -3\overset{''}{.}163\ 672 = -3°09'49''$$
$$\pi = +0\overset{''}{.}930\ 249 = 55'48''.9$$

The *Astronomical Ephemeris* gives the following values :

$$\lambda = 113°39'28\overset{''}{.}27$$
$$\beta = -3°09'49\overset{''}{.}22$$
$$\pi = 55'48\overset{''}{.}985$$

Lower accuracy. - Of course, when no high accuracy is required, the calculation may be considerably simplified :

- unless T is large, the terms in T^2 and T^3 in the formulae for L', M, M', D and F may be dropped ;

- Ω is not needed ;

- drop the additive terms to L', M, M', D and F ;

- use only a limited number of periodic terms in the expressions for λ, B and π ;

- put $\beta = B$.

As an exercise, calculate the coordinates of the Moon for 1979 December 7 at 0^h ET, with the above-mentioned simplifications. Compare your results with those of Example 30.a.

Angular speed of the Moon

For some applications it may be interesting to have a good value of the Moon's angular speed. Of course, this value could be deduced from two or more calculated positions of the Moon. However, if an accuracy of approximately 0.005 degree/day is sufficient, the speed can be directly calculated by means of the following formula.

For the given instant (ET !), calculate the JD (see Chapter 3), and then T by means of formula (15.1). Then calculate the angles M, M', D and F by means of the formulae given on page 148. Then the *geocentric* angular motion of the Moon in *ecliptical longitude*, expressed in degrees per day, is

```
    13.176 397
  + 1.434 006 cos M'
  + 0.280 135 cos 2D
  + 0.251 632 cos (2D - M')
  + 0.097 420 cos 2M'
  - 0.052 799 cos 2F
  + 0.034 848 cos (2D + M')
  + 0.018 732 cos (2D - M)
  + 0.010 316 cos (2D - M - M')
  + 0.008 649 cos (M - M')
  - 0.008 642 cos (2F + M')
  - 0.007 471 cos (M + M')
  - 0.007 387 cos D
  + 0.006 864 cos 3M'
  + 0.006 650 cos (4D - M')
  + 0.003 523 cos (2D + 2M')
  + 0.003 377 cos (4D - 2M')
  + 0.003 287 cos 4D
  - 0.003 193 cos M
  - 0.003 003 cos (2D + M)
  + 0.002 577 cos (M' - M + 2D)
  - 0.002 567 cos (2F - M')
  - 0.001 794 cos (2D - 2M')
  - 0.001 716 cos (M' - 2F - 2D)
  - 0.001 698 cos (2D + M - M')
  - 0.001 415 cos (2D + 2F)
  + 0.001 183 cos (2M' - M)
  + 0.001 150 cos (D + M)
  - 0.001 035 cos (D + M')
  - 0.001 019 cos (2F + 2M')
  - 0.001 006 cos (M + 2M')
```

In order to obtain the Moon's speed with respect to the moving Sun, change 13.176 397 into 12.190 749, and the coefficient of cos M to −0.036 211.

31

ILLUMINATED FRACTION OF THE MOON'S DISK

The illuminated fraction k of the Moon's disk, as seen from the center of the Earth, can be calculated from

$$k = \frac{1 + \cos i}{2} \qquad (31.1)$$

where i is the Moon's phase angle, that is the angular distance Sun - Earth as seen from the Moon.

The phase angle i can be found as follows. Find the Sun's longitude $\odot$ (Chapter 18), and the Moon's longitude λ and latitude β (Chapter 30). For the Moon, it is sufficient to take into account only a small number of periodic terms. For its latitude, for instance, it is sufficient to calculate

$$
\begin{aligned}
\beta = \ &+5°1282 \sin F \\
&+0°2806 \sin (F + M') \\
&+0°2777 \sin (M' - F) \\
&+0°1732 \sin (2D - F)
\end{aligned}
$$

Then, calculate d from

$$\cos d = \cos (\lambda - \odot) \cos \beta \qquad (31.2)$$

d being between 0 and 180 degrees. Then we have, with sufficient accuracy,

$$i = 180° - d - 0°1468 \frac{1 - 0.0549 \sin M'}{1 - 0.0167 \sin M} \sin d \qquad (31.3)$$

where M and M' are, as before, the mean anomalies of Sun and Moon, respectively.

Example 31.a : Calculate the illuminated fraction of the Moon's disk on 1979 December 25 at 0^h ET.

Instead of calculating the coordinates of Sun and Moon ourselves, we take them from the *A.E.* :

$$\odot = 272°35'23''$$
$$\lambda = 346°39'01''$$
$$\beta = -1°22'54''$$

whence, by means of formula (31.2), $d = 74°065$.
We have further

$$JD = 2444\ 232.5 \qquad T = +0.799\ 794\ 6612$$

whence, from the expressions given in Chapter 30,

$$M = 350°32 \qquad M' = 357°20$$

and then, by formulae (31.3) and (31.1),

$$i = 180° - 74°065 - \left(0°1468 \times \frac{1.0027}{1.0028} \times \sin 74°065\right)$$

$$i = 180° - 74°065 - 0°141 = 105°794$$

$$k = 0.3639, \quad \text{which should be rounded to 0.36.}$$

Lower accuracy, though still a good result, is obtained by neglecting the Moon's latitude and by calculating an approximate value of i as follows :

$$
\begin{aligned}
i = 180° - D &- 6°289 \sin M' \\
&+ 2°100 \sin M \\
&- 1°274 \sin (2D - M') \\
&- 0°658 \sin 2D \\
&- 0°214 \sin 2M' \\
&- 0°112 \sin D
\end{aligned}
\qquad (31.4)
$$

Example 31.b : Calculate again the illuminated fraction of the Moon's disk for 1979 December 25.0, but now using formula (31.4).

We have

$$JD = 2444\ 232.5 \qquad T = +0.799\ 794\ 6612$$

whence, from the expressions given in Chapter 30,

$$M = 350°324 \qquad M' = 357°202 \qquad D = 72°997$$

Then, by formula (31.4), $i = 105°843$

156

and, by formula (31.1), $k = 0.3635$, which again rounds to 0.36.

As an exercise, calculate the illuminated fraction of the Moon's disk for 0^h ET of the following dates, and compare your result with the value given in the *Astronomical Ephemeris* :

	A.E.
1978 October 24	0.50
1978 December 13	0.98
1979 April 1	0.18
1979 December 9	0.73

For 1979 December 9.0, the Soviet almanac *Astronomicheskii Ezhe-godnik* gives $k = 0.74$ instead of 0.73. Who is correct ?

32

Phases of the Moon

The times of the *mean* phases of the Moon, already affected by the Sun's aberration, are given by

$$JD = 2415\,020.759\,33 + 29.530\,588\,68\,k$$
$$+ 0.000\,1178\,T^2 \tag{32.1}$$
$$- 0.000\,000\,155\,T^3$$
$$+ 0.000\,33\,\sin(166°56 + 132°87\,T - 0°009\,173\,T^2)$$

These instants are expressed in Ephemeris Time (Julian Ephemeris Days). In the formula above, an integer value of k gives a New Moon, an integer value increased by

> 0.25 gives a First Quarter,
> 0.50 gives a Full Moon,
> 0.75 gives a Last Quarter

Any other value for k will give meaningless results !

A negative value of k gives a lunar phase before the year 1900, while k is positive after the beginning of the year 1900. Thus, for example,

> +479.00 and −2793.00 correspond to a New Moon,
> +479.25 and −2792.75 correspond to a First Quarter,
> +479.50 and −2792.50 correspond to a Full Moon,
> +479.75 and −2792.25 correspond to a Last Quarter.

An approximate value of k is given by

$$k \simeq (\text{year} - 1900) \times 12.3685 \tag{32.2}$$

where the "year" should be taken with decimals, for example 1977.25 for the end of March 1977.

Finally, in formula (32.1) T is the time in Julian centuries from 1900 January 0.5. Once the correct value of k has been found,

T can be calculated with a sufficient accuracy from

$$T = \frac{k}{1236.85} \qquad (32.3)$$

Then calculate the following angles, which are expressed in degrees and decimals and may be reduced to the interval $0 - 360$ degrees before calculating further.

Sun's mean anomaly at time JD :

$$M = 359.2242 + 29.105\,356\,08\,k$$
$$- 0.000\,0333\,T^2$$
$$- 0.000\,003\,47\,T^3$$

Moon's mean anomaly :

$$M' = 306.0253 + 385.816\,918\,06\,k$$
$$+ 0.010\,7306\,T^2$$
$$+ 0.000\,012\,36\,T^3$$

Moon's argument of latitude :

$$F = 21.2964 + 390.670\,506\,46\,k$$
$$- 0.001\,6528\,T^2$$
$$- 0.000\,002\,39\,T^3$$

To obtain the time of the *true* phase, the following corrections should be added to the time of the mean phase given by (32.1). The following coefficients are given in decimals of a day, and smaller quantities have been neglected.

For New and Full Moon :

$$
\begin{aligned}
&+ (0.1734 - 0.000\,393\,T)\,\sin M\\
&+ 0.0021\,\sin 2M\\
&- 0.4068\,\sin M'\\
&+ 0.0161\,\sin 2M'\\
&- 0.0004\,\sin 3M'\\
&+ 0.0104\,\sin 2F\\
&- 0.0051\,\sin (M + M')\\
&- 0.0074\,\sin (M - M')\\
&+ 0.0004\,\sin (2F + M)\\
&- 0.0004\,\sin (2F - M)\\
&- 0.0006\,\sin (2F + M')\\
&+ 0.0010\,\sin (2F - M')\\
&+ 0.0005\,\sin (M + 2M')
\end{aligned}
\qquad (32.4)
$$

For First and Last Quarter :

$$+ (0.1721 - 0.0004\,T)\ \sin M$$
$$+ 0.0021 \sin 2M$$
$$- 0.6280 \sin M'$$
$$+ 0.0089 \sin 2M'$$
$$- 0.0004 \sin 3M'$$
$$+ 0.0079 \sin 2F$$
$$- 0.0119 \sin (M + M')$$
$$- 0.0047 \sin (M - M')$$
$$+ 0.0003 \sin (2F + M)$$
$$- 0.0004 \sin (2F - M)$$
$$- 0.0006 \sin (2F + M')$$
$$+ 0.0021 \sin (2F - M')$$
$$+ 0.0003 \sin (M + 2M')$$
$$+ 0.0004 \sin (M - 2M')$$
$$- 0.0003 \sin (2M + M')$$

$$(32.5)$$

and, in addition :

for First Quarter : $+ 0.0028 - 0.0004 \cos M + 0.0003 \cos M'$

for Last Quarter : $- 0.0028 + 0.0004 \cos M - 0.0003 \cos M'$

Example 32.a :   Calculate the instant of the New Moon occurring in February 1977.

Mid-February 1977 being equal to 1977.13, we find by means of formula (32.2)

$$k \simeq (1977.13 - 1900) \times 12.3685 = 953.982$$

whence $k = 954$, since k should be an integer for the New Moon phase. Then, by formula (32.3), $T = +0.77131$, and then formula (32.1) gives

$$JD = 2443\ 192.9407$$

With $k = 954$ and $T = +0.77131$, we further find

$$M = 28125°7339 = 45°7339$$
$$M' = 368375°3715 = 95°3715$$
$$F = 372720°9585 = 120°9585$$

The correcting terms given by (32.4) are then, writing extra decimals :

```
+0.123956      +0.005639
+0.002099      -0.000381
-0.405014      +0.000111
-0.003001      +0.000232
+0.000384      +0.000551
-0.009176      -0.000417
-0.003202
```

the sum of which is -0.2882 day. Thus, the time of the true New Moon is

$$JD = 2443\,192.9407 - 0.2882 = 2443\,192.6525$$

which corresponds to 1977 February 18.1525

= 1977 February 18 at $3^h39^m\!.6$ ET.

The correct value, deduced from the data of the $A.E.$, is $3^h37^m\!.6$ ET.

Example 32.b :   Calculate the time of the Last Quarter of November 1952.

Using the value year = 1952.88, formula (32.2) gives $k \simeq 654.05$, and thus we take $k = 653.75$. We then find

$$T = +0.52856 \qquad JD = 2434\,326.3814$$

$$M = 306^\circ\!.8507$$
$$M' = 173^\circ\!.8385$$
$$F = 182^\circ\!.1395$$

and the total correction given by (32.5) is -0.2261 day, whence

$$JD = 2434\,326.3814 - 0.2261 = 2434\,326.1553$$

which corresponds to 1952 November 9.6553

= 1952 November 9 at $15^h43^m\!.6$ ET

or 15^h43^m UT, because in 1952 the difference ET - UT was +0.5 minute (see Chapter 5).

The correct value is indeed 15^h43^m UT.

Using the method described in this Chapter, the author has calculated all lunar phases of the years 1971 - 1975. It was found that no instant was more than 2 minutes in error. In 3/4 of the cases, the error was even less than 1.0 minute.

If an accuracy of half an hour is sufficient, one may drop the last term of formula (32.1), and the terms with coefficients less than 0.0030 in formulae (32.4) and (32.5).

33

Eclipses

Without too much calculation, it is possible to obtain with good accuracy the principal characteristics of an eclipse of Sun or Moon. For a solar eclipse, the situation is complicated by the fact that the phases of the event are different for different observers at the Earth's surface, while in the case of a lunar eclipse all observers see the same phase at the same instant.

For this reason, we will not consider here the calculation of the local circumstances of a solar eclipse. The interested reader may calculate these circumstances from the Besselian Elements published yearly in the *Astronomical Ephemeris* (renamed *Astronomical Almanac* since 1981), where he will find the necessary formulae. More formulae, together with numerical examples, are given in the *Explanatory Supplement to the A.E.* (out of print in 1984).

Firstly, calculate the time (JD) of the *mean* New or Full Moon, using formulae (32.1) to (32.3) of the preceding Chapter. Remember that k must be an integer for a New Moon (solar eclipse), and an integer increased by 0.5 for a Full Moon (lunar eclipse).

Then, calculate the values of M, M' and F for that instant, using the expressions given after formula (32.3).

The value of F will give a first information about the occurrence of a solar or lunar eclipse. If F differs from the nearest multiple of 180° by less than 13°.9, then there is certainly an eclipse ; if the difference is larger than 21°.0, there is no eclipse ; between these two values, the eclipse is uncertain and the case must be examined further. On a programmable calculating machine, use can be made of the following rule : there is no eclipse if $|\sin F| > 0.36$.

Note that, after one lunation, F increases by 30°.6705.

If F is near $0°$ or $360°$, the eclipse occurs near the Moon's ascending node. If F is near $180°$, the eclipse takes place near the descending node of the Moon's orbit.

To obtain the *time of maximum eclipse* (for the Earth generally in the case of a solar eclipse), the following corrections should be added to the time of mean conjunction given by (32.1). The following coefficients are given in decimals of a day, and smaller quantities have been neglected.

$$
\begin{aligned}
&+ (0.1734 - 0.000\ 393\ T)\ \sin M \\
&+ 0.0021\ \sin 2M \\
&- 0.4068\ \sin M' \\
&+ 0.0161\ \sin 2M' \\
&- 0.0051\ \sin (M + M') \\
&- 0.0074\ \sin (M - M') \\
&- 0.0104\ \sin 2F
\end{aligned}
\qquad (33.1)
$$

Note that the coefficient of $\sin 2F$ is negative here, while it was positive in (32.4); the reason is that we calculate here the time of greatest eclipse, not the time of conjunction in longitude.

Then calculate further :

$$
\begin{aligned}
S = \ &5.19\ 595 & C = \ &+\ 0.2070\ \sin M \\
&-\ 0.0048\ \cos M & &+\ 0.0024\ \sin 2M \\
&+\ 0.0020\ \cos 2M & &-\ 0.0390\ \sin M' \\
&-\ 0.3283\ \cos M' & &+\ 0.0115\ \sin 2M' \\
&-\ 0.0060\ \cos (M + M') & &-\ 0.0073\ \sin (M + M') \\
&+\ 0.0041\ \cos (M - M') & &-\ 0.0067\ \sin (M - M') \\
& & &+\ 0.0117\ \sin 2F
\end{aligned}
$$

$$
\gamma = S \sin F + C \cos F
$$

$$
\begin{aligned}
u = \ &0.0059 \\
&+\ 0.0046\ \cos M \\
&-\ 0.0182\ \cos M' \\
&+\ 0.0004\ \cos 2\ M' \\
&-\ 0.0005\ \cos (M + M')
\end{aligned}
$$

SOLAR ECLIPSES

In the case of a solar eclipse, γ represents the least distance from the axis of the Moon's shadow to the center of the Earth, in units of the equatorial radius of the Earth. The quantity γ is positive or negative depending upon the axis of the shadow passing north or south of the Earth's center. When γ is between +0.9972 and −0.9972, the solar eclipse is central : there exists a line of central eclipse on the Earth's surface.

The quantity u denotes the radius of the Moon's *umbral* cone in the fundamental plane, again in units of the Earth's equatorial radius. (The fundamental plane is the plane through the center of the Earth and perpendicular to the axis of the Moon's shadow). The radius of the *penumbral* cone in the fundamental plane is

$$u + 0.5460$$

If $|\gamma|$ is between 0.9972 and 1.5432 + u, the eclipse is not central. In most cases, it is then a partial one. However, when $|\gamma|$ is between 0.9972 and 1.0260, a part of the umbral cone may touch the surface of the Earth (within the polar regions), but the axis of the cone does *not* touch the Earth. These non-central total or annular eclipses occur when $0.9972 < |\gamma| < 0.9972 + |u|$. Between the years 1950 and 2100, there are seven eclipses of this type :

1950 March 18	annular, not central
1957 April 30	annular, not central
1957 October 23	total, not central
1967 November 2	total, not central
2014 April 29	annular, not central
2043 April 9	total, not central
2043 October 3	annular, not central

If $|\gamma| > 1.5432 + u$, no eclipse is visible from the Earth's surface.

In the case of a *central* eclipse, the type of the eclipse may be determined by the following rules :

if $u < 0$, the eclipse is total ;

if $u > +0.0047$, the eclipse is annular ;

if u is between 0 and +0.0047, the eclipse is either annular or annular-total.

In the last case, the ambiguity is removed as follows. Calculate

$$\omega = 0.00464 \cos W, \qquad \text{where} \qquad \sin W = \gamma$$

Then, if $u < \omega$, the eclipse is annular-total ; otherwise it is an annular one.

In the case of a *partial* solar eclipse, the greatest magnitude is attained at the point of the surface of the Earth which comes closest to the axis of shadow. The magnitude of the eclipse at that point is

$$\frac{1.5432 + u - |\gamma|}{0.5460 + 2u} \tag{33.2}$$

LUNAR ECLIPSES

In the case of a lunar eclipse, γ represents the least distance from the center of the Moon to the axis of the Earth's shadow, in units of the Earth's equatorial radius. The quantity γ is positive or negative depending upon the Moon's center passing north or south of the axis of shadow.

The radii at the distance of the Moon are :

for the penumbra : $\rho = 1.2847 + u$

for the umbra : $\sigma = 0.7404 - u$

while the magnitude of the eclipse may be found as follows :

for penumbral eclipses : $\dfrac{1.5572 + u - |\gamma|}{0.5450}$ $\qquad(33.3)$

for umbral eclipses : $\dfrac{1.0129 - u - |\gamma|}{0.5450}$ $\qquad(33.4)$

If the magnitude is negative, this indicates that there is no eclipse.

The *semidurations* of the partial and total phases in the *umbra* can be found as follows. Calculate

$$P = 1.0129 - u$$
$$T = 0.4679 - u$$
$$n = 0.5458 + 0.0400 \cos M'$$

Then the semidurations in *minutes* are :

partial phase : $\dfrac{60}{n} \sqrt{P^2 - \gamma^2}$ $\qquad$ total phase : $\dfrac{60}{n} \sqrt{T^2 - \gamma^2}$

Example 33.a :   Solar eclipse of 1978 October 2.

Since October 2 is the 275th day of the year, the given date corresponds to 1978.75. Formula (32.2) then gives

$$k \simeq 974.02, \quad \text{whence} \quad k = 974.$$

Then, by means of formulae (32.3) and (32.1),

$$JD = 2443\,783.5524$$

We find further

$$M = 267°8410 \qquad M' = 251°7102 \qquad F = 14°3687$$

Because F is between 13°9 and 21°0, the eclipse is uncertain. We find further :

$$S = 5.3067 \qquad C = -0.1616 \qquad \gamma = +1.1604 \qquad u = +0.0116$$

Because $|\gamma|$ is between 0.9972 and 1.5432 + u, the eclipse is a partial one. Using formula (33.2), we find that the maximum magnitude is

$$\frac{1.5432 + 0.0116 - 1.1604}{0.5460 + 0.0232} = 0.693$$

Because F is near 0°, the eclipse occurs at the Moon's ascending node. Because γ is positive, the eclipse is visible in the northern hemisphere of the Earth.

To obtain the time of maximum eclipse, we add to JD the terms given by formula (33.1). This gives

$$
\begin{aligned}
JD = 2443\,783.5524 \\
- 0.1730 \\
+ 0.0002 \\
+ 0.3862 \\
+ 0.0096 \\
- 0.0018 \\
- 0.0021 \\
- 0.0050
\end{aligned}
$$

which gives JD = 2443 783.767, corresponding to 1978 October 2 at 6^h24^m ET.

The correct values, given in the *A.E.*, are $6^h28^m\!.7$ ET and a maximum magnitude of 0.691.

Example 33.b : Solar eclipse of 1980 February 16.

As in the preceding Example, we find :

$$k = 991$$
$$JD = 2444\,285.572$$
$$M = 42°6321$$
$$M' = 330°5979$$
$$F = 175°7671$$

Corrected JD = 2444 285.871 = 1980 February 16 at 8^h54^m ET

$S = +4.9020$ $C = +0.1421$ $\gamma = +0.2201$ $u = -0.0069$

   Because $|\gamma| < 0.9972$, the eclipse is a central one. Because u is negative, the eclipse is a total one. Because $|\gamma|$ is small, the eclipse is visible from the equatorial regions of the Earth. The eclipse takes place near the descending node of the Moon's orbit, because $F \simeq 180°$.

Example 33.c : Lunar eclipse of June 1973.

We find successively :

$$k = 908.5$$
$$JD = 2441\,849.299$$
$$M = 161°4402$$
$$M' = 180°7011$$
$$F = 345°4506$$

Corrected JD = 2441 849.367 = 1973 June 15 at 20^h48^m ET

$S = +5.5285$ $C = +0.0640$ $\gamma = -1.3269$ $u = +0.0197$

   The eclipse took place near the Moon's ascending node ($F \simeq 360°$) and the Moon's center passed south of the center of the Earth's umbra ($\gamma < 0$).

   According to formula (33.4), the magnitude in the umbra was −0.612. Since this is negative, there was no eclipse in the umbra. Using formula (33.3), we find that the magnitude in the penumbra was 0.459. Thus the eclipse was a penumbral one.

   According to the *Connaissance des Temps*, maximum eclipse took place at $20^h50^m.7$ ET, and the magnitude was 0.469.

Example 33.d : Find the first lunar eclipse after 1978 July 1.

For 1978.5, formula (32.2) gives $k \simeq 970.93$. Thus we must try the value $k = 971.5$. This gives $F = 117°6924$, which differs more than 21 degrees from the nearest multiple of 180°, and thus gives no eclipse.

The next Full Moon, $k = 972.5$, gives $F = 148°3629$, hence again no eclipse. But it is evident that the next Full Moon will give $F \simeq 179°$ and thus give rise to an eclipse. We then find, as before :

$$k = 973.5$$
$$JD = 2443\,768.787$$
$$M = 253°2883$$
$$M' = 58°8017$$
$$F = 179°0334$$

Corrected JD = 2443 768.295 = 1978 September 16 at 19^h05^m ET
$$= 19^h04^m \text{ UT}$$

$S = +5.0176$ $C = -0.2134$ $\gamma = +0.2980$ $u = -0.0054$

Formula (33.4) then yields a magnitude of 1.332. Thus the eclipse is a total one in the umbra. We find further :

$$P = 1.0183 \qquad T = 0.4733 \qquad n = 0.5665$$

Semiduration of partial phase :

$$\frac{60}{0.5665} \sqrt{(1.0183)^2 - (0.2980)^2} = 103 \text{ minutes}$$

Semiduration of total phase :

$$\frac{60}{0.5665} \sqrt{(0.4733)^2 - (0.2980)^2} = 39 \text{ minutes}$$

Hence, in Universal Time :

beginning of partial phase :	19^h04^m	$- 103^m$	=	17^h21^m
beginning of total phase :	19^h04^m	$- 39^m$	=	18 25
maximum of the eclipse :				19 04
end of total phase :	19^h04^m	$+ 39^m$	=	19 43
end of partial phase :	19^h04^m	$+ 103^m$	=	20 47

Exercises

Find the first solar eclipse of the year 1979, and show that it was a total one visible from the northern hemisphere.

Was the solar eclipse of April 1977 a total or an annular one ?

Show that there was no eclipse of the Sun in July 1947.

Show that there will be four solar eclipses in the year 2000, and that all four will be partial eclipses.

Show that there was no lunar eclipse in January 1971.

Show that there were three total eclipses of the Moon in 1982.

Find the first lunar eclipse of the year 1234. (Answer : the partial lunar eclipse of 1234 March 17).

34

Illuminated Fraction
of the Disk of a Planet

As for the Moon (see Chapter 31), the illuminated fraction k of the disk of a planet, as seen from the Earth, can be calculated from

$$k = \frac{1 + \cos i}{2}$$

where i is the phase angle. In the case of a planet, this angle can be found from

$$\cos i = \frac{r^2 + \Delta^2 - R^2}{2\,r\,\Delta}$$

r being the planet's distance to the Sun, Δ its distance to the Earth, and R the distance Sun - Earth, all in astronomical units. Combining these two formulae, we find

$$k = \frac{(r + \Delta)^2 - R^2}{4\,r\,\Delta} \qquad (34.1)$$

If the planet's position has been obtained by the first method of Chapter 25, we can find k as follows :

$$k = \frac{r + \Delta + R \cos b \cos (l - \Theta)}{2\,\Delta}$$

Example 34.a : Find the illuminated fraction of the disks of Mercury, Venus and Mars on 1979 April 17 at 0^h ET.

We will use formula (34.1), and take the values of r, Δ and R from the *Astronomical Ephemeris*.

	Mercury	Venus	Mars
r	0.466 674	0.728 149	1.387 513
Δ	0.785 473	1.300 500	2.300 530
R	1.003 712	1.003 712	1.003 712
k	0.382	0.821	0.986

For Mercury and Venus, k can take all values between 0 and 1. For Mars, k can never be less than approximately 0.838. In the case of Jupiter, i is never less than 12°, whence k can vary only between 0.989 and 1.

In the case of *Venus*, an *approximate* value of k can be found as follows.
Calculate T by means of formula (18.1). Then,

$$V = 63°07 + 22518°443\ T$$
$$M = 178°48 + 35999°050\ T$$
$$M' = 212°60 + 58517°804\ T$$
$$W = V + 1°92 \sin M + 0°78 \sin M' \tag{34.2}$$
$$\Delta^2 = 1.523\ 209 + 1.446\ 664 \cos W \qquad (\Delta > 0)$$
$$k = \frac{(0.723\ 332 + \Delta)^2 - 1}{2.893\ 329\ \Delta}$$

Example 34.b : Find the illuminated fraction of the disk of Venus on 1979 April 17.0 ET, using the approximate method described above.

We find successively :

JD = 2443 980.5 $W = V - 1°88 + 0°12 = 276°08$
T = +0.792 895 277 Δ^2 = 1.676 435
V = 17917°84 = 277°84 Δ = 1.294 772
M = 28721°96 = 281°96
M' = 46611°09 = 171°09 k = 0.820

The correct value, found in the preceding example, is 0.821.

The *elongation* ψ of a planet can be calculated from formula (25.11). If the distances R, r and Δ are known, it can also be found from

$$\cos \psi = \frac{R^2 + \Delta^2 - r^2}{2 R \Delta} \tag{34.3}$$

In the case of *Venus*, an *approximate* value of ψ can be found by first calculating Δ from (34.2), and then

$$\cos \psi = \frac{\Delta^2 + 0.4768}{2 \Delta} \tag{34.4}$$

Taking for Venus the values given in Example 34.a, we find, by formula (34.3), $\cos \psi = 0.830\ 649$, whence $\psi = 33°50'$.

Taking the approximate value of Δ found in Example 34.b, formula (34.4) gives $\cos \psi = 0.8315$, whence $\psi = 33°45'$.

35

CENTRAL MERIDIAN OF JUPITER

For Jupiter three rotational systems have been adopted. System I applies for the equatorial regions of the cloud cover of the planet, System II for the zones which are farther north or south from the equator, while System III applies to radio emissions of Jupiter. In this Chapter we will consider only Systems I and II, which are of interest for the visual observers.

The longitude of the central meridian of Jupiter, for each day of the year at 0^h UT, can be found in several astronomical almanacs. However, this longitude can also be calculated, for any instant, by means of the following method. The accuracy of the result is approximately 0.1 degree, which is sufficient in many cases. We take into account the eccentricities of the orbits of Jupiter and the Earth, and the effects of the planet's phase and light-time (the time needed for the light to reach the Earth from the planet). A long-period term in the motion of Jupiter too is taken into account, but all other periodic terms in the motions of Earth and Jupiter have been neglected. For the eccentricities of the orbits, their values for the year 1980 have been used.

For the given instant (ET !), calculate the JD (see Chapter 3), and then proceed as follows :

Number of days (and decimals of a day) since 1899 December 31 at 12^h ET) :

$$d = \text{JD} - 2415\,020$$

Argument for the long-period term in the motion of Jupiter :

$$V = 134.63 + 0.001\,115\,87\,d$$

Mean anomalies of Earth and Jupiter :

$$M = 358.476 + 0.985\,6003\,d$$

$$N = 225.328 + 0.083\,0853\,d + 0.33\,\sin V$$

Difference between the mean heliocentric longitudes of Earth and Jupiter :

$$J = 221.647 + 0.902\,5179\,d - 0.33\,\sin V$$

It should be noted that the angles V, M, N and J are expressed in degrees and decimals. If necessary, they should be reduced to the interval 0 - 360 degrees ; this depends on your calculator.

Equations of the center of Earth and Jupiter, in degrees :

$$A = 1.916\,\sin M + 0.020\,\sin 2M$$

$$B = 5.552\,\sin N + 0.167\,\sin 2N$$

and then

$$K = J + A - B$$

Radius vector of the Earth :

$$R = 1.00014 - 0.01672\,\cos M - 0.00014\,\cos 2M$$

Radius vector of Jupiter :

$$r = 5.20867 - 0.25192\,\cos N - 0.00610\,\cos 2N$$

Distance Earth - Jupiter :

$$\Delta = \sqrt{r^2 + R^2 - 2\,r\,R\,\cos K}$$

The distances R, r and Δ are expressed in astronomical units, and Δ should of course be taken positive. The phase angle of Jupiter (that is, the angle Earth-Jupiter-Sun) is then given by

$$\sin \psi = \frac{R}{\Delta}\,\sin K$$

The angle ψ always lies between $-12°$ and $+12°$. Because R and Δ are always positive, the angle ψ has the same sign as $\sin K$.

The longitude of the central meridian of Jupiter is then

in System I :

$$\lambda_1 = 268°.28 + 877°.816\,9088 \left(d - \frac{\Delta}{173} \right) + \psi - B$$

in System II :

$$\lambda_2 = 290°.28 + 870°.186\,9088 \left(d - \frac{\Delta}{173} \right) + \psi - B$$

where $-\Delta/173$ is the correction for the light-time, expressed in days. The denominator 173 results from the fact that the light-time for unit distance is 1/173 day.

The thus obtained values for λ_1 and λ_2 should be reduced to the interval $0° - 360°$, by adding or subtracting a convenient multiple of 360 degrees. Moreover, it should be noted that the results refer to the geometric (the "true") disk of Jupiter. Jupiter actually has a very small phase, and the longitudes of the "central meridian" of the illuminated disk can be obtained by adding to λ_1 and λ_2 the *correction for phase* which is equal to

$$\pm\ 57°.3\ \sin^2\frac{\psi}{2}$$

and the sign is opposite the sign of $\sin K$.

As mentioned on page 175, the results are valid for an instant expressed in Ephemeris Time. If λ_1 and λ_2 are required for an instant expressed in UT, then that instant should first be converted to ET by adding the quantity ΔT (see Chapter 5). However, another method is calculating the longitudes for an ET which is *numerically* equal to the given instant UT, and then correcting λ_1 and λ_2 as follows :

$$\text{correction to } \lambda_1\ :\quad C_1 = +0°.01016\ \Delta T$$

$$\text{correction to } \lambda_2\ :\quad C_2 = +0°.01007\ \Delta T$$

where ΔT is the difference $ET - UT$ expressed in *seconds of time*. In this manner, we can calculate λ_1 and λ_2 for a given date at 0^h UT by first calculating them for 0^h ET of that date, and then adding the corrections C_1 and C_2 , as in Example 35.a.

Example 35.a :   Calculate λ_1 and λ_2 for 1980 June 30 at 0^h UT.

This date corresponds to JD 2444 420.5, and we find successively :

$d = 29\,400.5$	$A = +0°.143$	$\sin \psi = +0.15817$
$V = 167°.44$	$B = +2°.780$	$\psi = +9°.101$
$M = 29335°.618 = 175°.618$	$K = 113°.416$	$d - \Delta/173 = 29400.46591$
$N = 2668°.149 = 148°.149$	$R = 1.01667$	$\Delta T = +51$ seconds
$J = 26756°.053 = 116°.053$	$r = 5.41995$	$C_1 = +0°.52$
	$\Delta = 5.89823$	$C_2 = +0°.51$

From this we deduce, for the geometric disk of Jupiter :

at 0^h ET : $\lambda_1 = 25\,808\,500°.70 = 100°.70$
$\lambda_2 = 25\,584\,197°.15 = 77°.15$

at 0^h UT : $\lambda_1 = 100°.70 + 0°.52 = 101°.22$
$\lambda_2 = 77°.15 + 0°.51 = 77°.66$

The *Astronomical Ephemeris* gives $\lambda_1 = 101°.30$ and $\lambda_2 = 77°.74$.

Finally, we find that the correction for the illuminated disk is $-0°.36$, exactly the value given by the *A.E.*

Let D_e be the planetocentric angular distance of the Earth from the equator of Jupiter ; this is the same as the planetocentric declination of the Earth, or as the planetocentric latitude of the center of the planet's disk as seen from the Earth. Similarly, let us denote by D_s the planetocentric declination of the Sun.

Since the inclination of the equator of Jupiter to the plane of the planet's orbit is $3°.07$, the extreme values of D_s are $+3°.07$ and $-3°.07$. The extreme values of D_e are $+3°.4$ and $-3°.4$.

For any given instant, the values of D_s and D_e can be obtained as follows. Because these values vary very slowly with time, it's not necessary to make a distinction between ET and UT in this case.

Calculate, as explained before, the values of d, V, M, N, J, A, B, K, R, r, Δ and ψ. Then find Jupiter's heliocentric longitude λ referred to the equinox of the year 1900 by the formula

$$\lambda = 238°.05 + 0°.083\,091\,d + 0°.33 \sin V + B$$

Then we obtain, in degrees and decimals,

$$D_s = 3.07 \sin (\lambda + 44°.5)$$

$$D_e = D_s - 2.15 \sin \psi \cos (\lambda + 24°) - 1.31 \frac{r - \Delta}{\Delta} \sin (\lambda - 99°.4)$$

Calculated in this manner, D_s will rarely be more than $0°.01$ in error, and the error in D_e will rarely exceed $0°.03$.

Example 35.b : Let us take the same instant as in Example 35.a, where we have found the following values :

$$d = 29400.5 \qquad r = 5.41995$$
$$V = 167°.44 \qquad \Delta = 5.89823$$
$$B = +2°.780 \qquad \sin \psi = +0.15817$$

Using the above formulae, we then find

$$\lambda = 2683°.82 = 163°.82$$
$$D_s = -1°.46$$
$$D_e = -1°.46 + 0°.34 + 0°.10 = -1°.02$$

The *Astronomical Ephemeris* gives the same values for D_s and D_e.

36

Positions of the Satellites of Jupiter

With the following method it is possible to calculate, for any given instant, the positions of the four great satellites of Jupiter with respect to the planet, as seen from the Earth. The results are good, but not extremely accurate, and therefore may not be used for accurate calculations.

First, convert the date and the instant (ET) into the Julian Day, using the method described in Chapter 3. Then, obtain the following quantities as explained in Chapter 35 : d, V, M, N, J, A, B, K, R, r, Δ, ψ, and the planetocentric declination D_e of the Earth.

For each of the four satellites we now calculate an angle u which is measured from the inferior conjunction with Jupiter, so that $u = 0°$ corresponds to the satellite's inferior conjunction, $u = 90°$ with its greatest western elongation, $u = 180°$ with the superior conjunction, and $u = 270°$ with the greatest eastern elongation :

$$u_1 = 84°.5506 + 203°.405\ 8630 \left(d - \frac{\Delta}{173} \right) + \psi - B$$

$$u_2 = 41°.5015 + 101°.291\ 6323 \left(d - \frac{\Delta}{173} \right) + \psi - B$$

$$u_3 = 109°.9770 + 50°.234\ 5169 \left(d - \frac{\Delta}{173} \right) + \psi - B$$

$$u_4 = 176°.3586 + 21°.487\ 9802 \left(d - \frac{\Delta}{173} \right) + \psi - B$$

If necessary, these angles u should be reduced to the interval $0° - 360°$. In order to obtain more accurate values, the results can be corrected as follows. Calculate the angles G and H by means of the formulae

$$G = 187°.3 + 50°.310\ 674\ \left(d - \frac{\Delta}{173}\right)$$

$$H = 311°.1 + 21°.569\ 229\ \left(d - \frac{\Delta}{173}\right)$$

Then we have the following corrections, in degrees :

correction to u_1 :	$+0.472 \sin 2(u_1 - u_2)$
correction to u_2 :	$+1.073 \sin 2(u_2 - u_3)$
correction to u_3 :	$+0.174 \sin G$
correction to u_4 :	$+0.845 \sin H$

The first correction is due to a periodic perturbation of satellite I by satellite II. The second correction is a perturbation of II
by III. The two last corrections are due to the eccentricities of the
orbits of satellites III and IV. (The orbits of I and II are almost
exactly circular).

It should be noted that we take into account only the largest periodic terms in the motions of the satellites. There are many other
(but smaller) periodic terms. For instance, satellite I is perturbed
by satellite III too, satellite III by II and by IV, etc.

The distances of the satellites to the center of Jupiter, in units
of Jupiter's equatorial radius, are given by

$$r_1 = \quad 5.9061 - 0.0244 \cos 2(u_1 - u_2)$$

$$r_2 = \quad 9.3972 - 0.0889 \cos 2(u_2 - u_3)$$

$$r_3 = 14.9894 - 0.0227 \cos G$$

$$r_4 = 26.3649 - 0.1944 \cos H$$

where the *un*corrected values of u_1 etc. should be used. In these expressions, the periodic terms are again due to mutual perturbations
of the satellites or to their orbital eccentricities.

The apparent rectangular coordinates X and Y of the satellites
with respect to the center of the disk of Jupiter, and expressed in
units of Jupiter's equatorial radius, are then given by

$$X_1 = r_1 \sin u_1 \qquad \text{and} \qquad Y_1 = -r_1 \cos u_1 \sin D_e$$

with similar expressions for the other three satellites. X is measured positively to the west of Jupiter, negatively to the east, and
the X-axis coincides with the equator of the planet. Y is positive
to the north, negative to the south.

Example 36.a : Calculate the configuration of the satellites of Jupiter on 1980 June 30 at 0^h Ephemeris Time.

For this instant we have found, in Examples 35.a and 35.b,

$$d = 29400.5 \qquad \Delta = 5.89823$$
$$B = +2°.780 \qquad D_e = -1°.02$$
$$\psi = +9°.101 \qquad d - \Delta/173 = 29400.46591$$

By means of the formulae given in the present Chapter we then find successively :

$$u_1 = 5980\ 318°.013 = 358°.013 \qquad G = 1479\ 344°.6 = 104°.6$$
$$u_2 = 2978\ 069°.005 = 149°.005 \qquad H = 634\ 456°.5 = 136°.5$$
$$u_3 = 1477\ 034°.500 = 314°.500 \qquad 2\,(u_1 - u_2) = 418°.02 = 58°.02$$
$$u_4 = 631\ 939°.309 = 139°.309 \qquad 2\,(u_2 - u_3) = -330°.99 = 29°.01$$

correction to u_1 :	+0°.400	corrected u_1 = 358°.413
correction to u_2 :	+0°.520	corrected u_2 = 149°.525
correction to u_3 :	+0°.168	corrected u_3 = 314°.668
correction to u_4 :	+0°.582	corrected u_4 = 139°.891

(It's just a coincidence that all four corrections are positive).

$$r_1 = 5.9061 - 0.0129 = 5.8932 \qquad X_1 = -\ 0.163 \qquad Y_1 = +0.105$$
$$r_2 = 9.3972 - 0.0777 = 9.3195 \qquad X_2 = +\ 4.726 \qquad Y_2 = -0.143$$
$$r_3 = 14.9894 + 0.0057 = 14.9951 \qquad X_3 = -10.664 \qquad Y_3 = +0.188$$
$$r_4 = 26.3649 + 0.1410 = 26.5059 \qquad X_4 = +17.076 \qquad Y_4 = -0.361$$

With these values of X and Y we can draw the following figure which shows the configuration of the satellites at the given time. In this figure the South is up, and the West to the left, as in the field of an inverting telescope in the northern hemisphere.

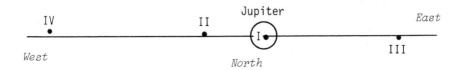

Satellite I is evidently in transit over the disk of Jupiter, since its distance to the planet's center is less than 1, and because u_1 = 358°.413, which is close to 360° (= 0°), indicating an *inferior* conjunction.

Mutual conjunctions. — Two satellites are in conjunction when their X-coordinates are equal. The difference between the Y-coordinates then corresponds to the separation of the satellites.

It should be noted, however, that the values of Y, calculated by means of the method described in the present Chapter, are less accurate than the values of the X-coordinates. Indeed, in our simplified method we have assumed that the satellites are situated exactly in the plane of Jupiter's equator. Actually, the four satellites can reach extreme latitudes of $0°02'$, $0°31'$, $0°19'$ and $0°44'$, respectively, with respect to the equatorial plane of the planet. As a consequence, mutual occultations cannot be calculated with certainty by means of the simplified method described in this Chapter. In the case of a very close conjunction, it's even not possible to deduce which of the two satellites passes to the north of the other.

Example 36.b :   Calculate the time of the conjunction between the satellites II and III on the morning of 1979 January 28.

We calculate the X and Y coordinates of these two satellites for several instants, at intervals of 0.04 day. The results are as follows.

1979, ET	X_2	X_3	Y_2	Y_3	$X_2 - X_3$
Jan 28.24	+4.279	+4.050	+0.080	+0.140	+0.229
28.28	+3.671	+3.539	+0.083	+0.141	+0.132
28.32	+3.044	+3.024	+0.086	+0.143	+0.020
28.36	+2.401	+2.505	+0.087	+0.144	-0.104
28.40	+1.746	+1.982	+0.089	+0.144	-0.236

By interpolation, we find that $X_2 = X_3$ at 7^h50^m4 ET, which corresponds to 7^h49^m6 UT since in 1979 the difference ET – UT was +50 seconds. Of course, the result should be rounded to the nearest minute, or 7^h50^m UT.

On page 60 of the *Handbook* of the British Astronomical Association for 1979, we read that an *occultation* of satellite III by satellite II was predicted for 1979 January 28, the time of the conjunction being 7^h56^m UT. This is six minutes later than the time we have just found. This difference cannot be surprising, however, because near the time of their conjunction the two satellites were moving at nearly the same apparent speed, as we can see from the values of X_2 and X_3. We further see that the difference between the values of Y_2 and Y_3 is small, which indicates a close conjunction between satellites II and III.

Conjunctions with Jupiter. - A satellite is in inferior conjunction with Jupiter when its X-coordinate is zero and passing from negative to positive ; or, what is the same, when the corresponding angle u, reduced to the interval $0° - 360°$, is $0°$ or $360°$.

Similarly, a satellite is in superior conjunction with Jupiter when its X-coordinate, passing from positive to negative, becomes zero, or when $u = 180°$.

Exercise. - On 1978 April 16, the satellites I and III were almost simultaneously in conjunction with Jupiter. Confirms this with your program, and compare your results with the following data taken from the *Astronomical Ephemeris* :

$$17^h 27^m \text{ UT} \quad \text{III in superior conjunction}$$
$$17 \quad 30 \qquad\qquad \text{I in inferior conjunction}$$

37

SEMIDIAMETERS OF SUN, MOON AND PLANETS

Sun and Planets

When not available directly from almanac data, the semidiameters s of the Sun and planets can be computed from

$$s = \frac{s_o}{\Delta}$$

where s_o is the body's semidiameter at unit distance (1 AU),
Δ is the body's distance to the Earth, in AU.

The following values of s_o should be used :

Sun	959″63		Saturn :	
Mercury	3.34		equatorial	83″33
			polar	74.57
Venus	8.41			
Mars	4.68		Uranus	34.28
Jupiter :			Neptune	36.56
equatorial	98.47			
polar	91.91			

Moon

When not available directly from almanac data, the semidiameter s' of the Moon, expressed in seconds of a degree (″), can be computed from

$$s' = \frac{56\,204.92}{D} = \frac{358\,482\,800}{D'} = 0.272\,476\;\pi'$$

where D = geocentric distance of the Moon in units of the equatorial radius of the Earth ;

D' = geocentric distance of the Moon in kilometers ;

π' = the horizonal equatorial parallax of the Moon in seconds of a degree ($''$).

Computed in this way, the semidiameter is geocentric, that is it applies to a fictitious observer located at the center of the Earth. The observed semidiameter of the Moon will be slightly larger than the geocentric diameter. It can be obtained, with sufficient accuracy for many purposes, by multiplying the geocentric value by

$$1 + \frac{\sin h}{D}$$

where h is the altitude of the Moon above the observer's horizon, D is, as above, the geocentric distance of the Moon in units of the Earth's equatorial radius.

The increase in the Moon's semidiameter, due to the fact that the observer is not geocentric, is zero when the Moon is on the horizon, and a maximum (between $14''$ and $18''$) when the Moon is at the zenith.

38

STELLAR MAGNITUDES

Adding stellar magnitudes

If two stars are of magnitudes m_1 and m_2 , respectively, their combined magnitude m can be calculated as follows :

$$x = 0.4 \, (m_2 - m_1)$$

$$m = m_2 - 2.5 \, \log \, (10^x + 1)$$

where the logarithm is in base 10.

Example 38.a : The magnitudes of the components of Castor (α Gem) are 1.96 and 2.89. Calculate the combined magnitude.

One finds

$$x = 0.4 \, (2.89 - 1.96) = 0.372$$

$$m = 2.89 - 2.5 \, \log \, (10^{0.372} + 1) = 1.58$$

Brightness ratio

If two stars are of magnitudes m_1 and m_2 , respectively, the ratio I_1/I_2 of their apparent luminosities can be calculated as follows :

$$x = 0.4 \, (m_2 - m_1) \qquad\qquad \frac{I_1}{I_2} = 10^x$$

If the brightness ratio I_1/I_2 is given, the corresponding magnitude difference $\Delta m = m_2 - m_1$ can be calculated from

$$\Delta m = 2.5 \, \log \, \frac{I_1}{I_2}$$

Example 38.b :   How many times is Vega (magnitude 0.14) brighter than Polaris (magnitude 2.12) ?

$$x = 0.4 \ (2.12 - 0.14) = 0.792$$

$$10^x = 6.19$$

Thus, Vega is 6.19 times as bright as the Pole Star.

Example 38.c :   A star is 500 times as bright as another one. The corresponding magnitude difference is

$$\Delta m = 2.5 \ \log 500 \ = \ 6.75$$

Distance and Absolute Magnitude

If π is a star's parallax expressed in seconds of a degree ($''$), this star's distance to us is equal to

$$\frac{1}{\pi} \ \text{parsecs} \qquad \text{or} \qquad \frac{3.2616}{\pi} \ \text{light-years}$$

If π is a star's parallax expressed in seconds of a degree ($''$), and m is the apparent magnitude of this star, its absolute magnitude M can be calculated from

$$M = m + 5 + 5 \ \log \pi$$

where, again, the logarithm is in base 10.

If d is the star's distance in parsecs, we have

$$M = m + 5 - 5 \ \log d$$

39

BINARY STARS

The orbital elements of a binary star are the following ones :

P = the period of revolution expressed in mean solar years ;

T = the time of periastron passage, generally given as a year
and decimals (for instance, 1945.62);

e = the eccentricity of the true orbit ;

a = the semimajor axis expressed in seconds of a degree ($''$) ;

i = the inclination of the plane of the true orbit to the plane
at right angles to the line of sight. For direct motion
in the apparent orbit, i ranges from 0° to 90° ; for
retrograde motion, i is between 90 and 180 degrees.
When i is 90°, the apparent orbit is a straight line
passing through the primary star ;

Ω = the position angle of the ascending node ;

ω = the longitude of periastron ; this is the angle in the plane
of the true orbit measured from the ascending node to
the periastron, taken always in the direction of motion.

When these orbital elements are known, the apparent position angle
θ and the angular distance ρ can be calculated for any given time
t, as follows.

$$n = \frac{360°}{P} \qquad\qquad M = n\ (t - T)$$

n is the mean annual motion of the companion, expressed in degrees
and decimals, and is always positive. M is the companion's mean
anomaly for the given time t.

Then solve Kepler's equation

$$E = M + e \sin E$$

by one of the methods described in Chapter 22, and then calculate the radius vector r and the true anomaly v from

$$r = a \ (1 - e \cos E)$$

$$\tan \frac{v}{2} = \sqrt{\frac{1 + e}{1 - e}} \ \tan \frac{E}{2}$$

Then find $(\theta - \Omega)$ from

$$\tan (\theta - \Omega) = \frac{\sin (v + \omega) \cos i}{\cos (v + \omega)} \qquad (39.1)$$

Of course, this equation can be written

$$\tan (\theta - \Omega) = \tan (v + \omega) \cos i$$

but in this case the correct quadrant for $(\theta - \Omega)$ is not determined. As in previous cases, it is better to leave equation (39.1) unchanged, and to apply to the numerator and the denominator of the fraction the conversion from rectangular to polar coordinates. This procedure will place the angle $(\theta - \Omega)$ at once in the correct quadrant.

When $(\theta - \Omega)$ is found, add Ω to obtain θ. If necessary, reduce the result to the interval $0° - 360°$.

The angular separation ρ is found from

$$\rho = \frac{r \cos (v + \omega)}{\cos (\theta - \Omega)}$$

Example 39.a : According to E. Silbernagel (1929), the orbital elements for η Coronae Borealis are :

P = 41.623 years	i = 59°025
T = 1934.008	Ω = 23°717
e = 0.2763	ω = 219°907
a = 0″907	

Calculate θ and ρ for the epoch 1980.0.

We find successively :

$$n = 8.64906$$
$$t - T = 1980.0 - 1934.008 = 45.992$$
$$M = 397°788 = 37°788$$
$$E = 49°897$$
$$r = 0″74557$$

$$v = 63°416$$

$$\tan(\theta - \Omega) = \frac{-0.500\ 813}{+0.230\ 440}$$

$$\theta - \Omega = -65°291$$

$$\theta = -41°574 = 318°4$$

$$\rho = 0''411$$

It is possible to write a program for obtaining a complete ephemeris of a binary for many years. For the HP-67 machine, the author has written such a program (150 steps). It contains as a subroutine the resolution of Kepler's equation, which can be used as an independent program. After entering the orbital elements, one enters the starting year of the required ephemeris, and the ephemeris interval in years. Then, without any push on a key, the machine displays successively, during pauses,

> the year,
> the position angle in degrees,
> the separation in ",

then the next year, and so on.

Try to write a similar program for your machine. As an exercise, calculate an ephemeris for γ Virginis, using the following elements (K. Strand, 1937) :

$P = 171.37$ years	$i = 146°05$
$T = 1836.433$	$\Omega = 31°78$
$e = 0.8808$	$\omega = 252°88$
$a = 3''746$	

Answer. – Here is an ephemeris with an interval of four years, starting with 1980. The position angle decreases with time, since i is between 90 and 180 degrees.

year =	θ =	ρ =
1980.0	296°86	3''90
1984.0	293.55	3.58
1988.0	289.55	3.23
1992.0	284.49	2.83
1996.0	277.62	2.37
2000.0	267.10	1.84
2004.0	246.26	1.18
2008.0	125.94	0.38
2012.0	24.65	1.35

Eccentricity of the apparent orbit

The apparent orbit of a binary star is an ellipse whose eccentricity e' is generally different from the eccentricity e of the true orbit. It may be interesting to know e', although this apparent eccentricity has no astrophysical significance.

The following formulae have been published in an article, by the author of this book, in the *Journal* of the British Astronomical Association, Vol. 89, pages 485-488, Aug. 1979.

$$A = (1 - e^2 \cos^2\omega) \cos^2 i$$

$$B = e^2 \sin \omega \cos \omega \cos i$$

$$C = 1 - e^2 \sin^2\omega$$

$$D = (A - C)^2 + 4B^2$$

$$e'^2 = \frac{2 \sqrt{D}}{A + C + \sqrt{D}}$$

It should be noted that e' is independent of the orbital elements a and Ω, and that is can be smaller as well as larger than the true eccentricity e.

Example 39.b :   Find the eccentricity of the apparent orbit of η Coronae Borealis. The orbital elements are given in Example 39.a.

We find

$$A = 0.25298$$
$$B = 0.01934$$
$$C = 0.96858$$
$$D = 0.51358$$

$$e' = 0.860$$

Hence, for this binary the apparent orbit is much more elongated than the true orbit.

40

LINEAR REGRESSION ; CORRELATION

In many cases, the result of a large number of observations is a series of points in a graph, each point being defined by a x-value and a y-value. It may be necessary to draw through the points the "best" fitting curve.

Several curves can be fitted through a series of points : a straight line, an exponential, a logarithmic curve, a polynomial, etc. We will consider here only the case of a straight line, a problem called *linear regression*.

We wish to calculate the coefficients of the linear equation

$$y = ax + b \tag{40.1}$$

using the least squares method. The slope a and the y-intercept b can be calculated by means of the formulae

$$a = \frac{n \, \Sigma xy - \Sigma x \, \Sigma y}{n \, \Sigma x^2 - (\Sigma x)^2}$$

$$b = \frac{\Sigma y \, \Sigma x^2 - \Sigma x \, \Sigma xy}{n \, \Sigma x^2 - (\Sigma x)^2} \tag{40.2}$$

where n is the number of points. The sign Σ indicates the summation. Thus, Σx signifies the sum of all the x-values, Σy the sum of all the y-values, Σx^2 the sum of the squares of all the x-values, Σxy the sum of the products xy of all the couples of values, etc. It should be noted that Σxy is not the same as $\Sigma x \times \Sigma y$ (the sum of the products is not the same as the product of the sums), and that $(\Sigma x)^2$ is not the same as Σx^2 (the square of the sum is not the same as the sum of the squares) !

An interesting astronomical application is to find the relation between the intrinsic brightness of a comet and its distance to the Sun. The apparent magnitude m of a comet can generally be represented by a formula of the form

$$m = g + 5 \log \Delta + \kappa \log r$$

which we already mentioned in Chapter 25. Here, Δ and r are the distance in astronomical units of the comet to the Earth and to the Sun, respectively. The absolute magnitude g and the coefficient κ must be deduced from the observations. This can be performed when the magnitude m has been measured during a sufficiently long period. For each value of m, the values of Δ and r must be deduced from the ephemeris, or calculated from the orbital elements.

In that case, the unknowns are g and κ. The above formula can be written

$$m - 5 \log \Delta = \kappa \log r + g$$

which is of the form (40.1), when we write $y = m - 5 \log \Delta$, and $x = \log r$. The quantity y may be called the "heliocentric" magnitude, because the effect of the variable distance to the Earth has been removed.

Example 40.a :   Table 40.A contains visual magnitude estimates m of the periodic comet Wild 2 (1978b), made by John Bortle. The corresponding values of r and Δ have been calculated from the orbital elements (IAUC 3177).

The quantities y and x are now used to calculate Σx, Σy, Σx^2, and Σxy. On a HP-67 and on some other machines, there exists a key ($\Sigma+$) whose use stores the different sums Σx, Σy^2, etc. in different registers. This key should be pressed each time after a couple of y and x values has been entered.

With the values of the table, we find

$$n = 19 \qquad \Sigma x = 4.2805 \qquad \Sigma x^2 = 1.0031$$
$$\Sigma y = 192.0400 \qquad \Sigma xy = 43.7943$$

whence, by formulae (40.2),

$$a = 13.67 \qquad b = 7.03$$

Consequently, the "best" line fitting the observations is

$$y = 13.67 x + 7.03$$
$$\text{or} \quad m - 5 \log \Delta = 13.67 \log r + 7.03$$

TABLE 40.A

1978, UT	m	r	Δ	$y =$ $m - 5 \log \Delta$	$x =$ $\log r$
Febr. 4.01	11.4	1.987	1.249	10.92	0.2982
5.00	11.5	1.981	1.252	11.01	0.2969
9.02	11.5	1.958	1.266	10.99	0.2918
10.02	11.3	1.952	1.270	10.78	0.2905
25.03	11.5	1.865	1.335	10.87	0.2707
March 7.07	11.5	1.809	1.382	10.80	0.2574
14.03	11.5	1.772	1.415	10.75	0.2485
30.05	11.0	1.693	1.487	10.14	0.2287
April 3.05	11.1	1.674	1.504	10.21	0.2238
10.06	10.9	1.643	1.532	9.97	0.2156
26.07	10.7	1.582	1.592	9.69	0.1992
May 1.08	10.6	1.566	1.610	9.57	0.1948
3.07	10.7	1.560	1.617	9.66	0.1931
8.07	10.7	1.545	1.634	9.63	0.1889
26.09	10.8	1.507	1.696	9.65	0.1781
28.09	10.6	1.504	1.703	9.44	0.1772
29.09	10.6	1.503	1.707	9.44	0.1770
June 2.10	10.5	1.498	1.721	9.32	0.1755
6.09	10.4	1.495	1.736	9.20	0.1746

Thus we have, for the periodic comet Wild 2 in 1978,

$$m = 7.03 + 5 \log \Delta + 13.67 \log r$$

Coefficient of Correlation

A correlation coefficient is a statistical measure of the degree to which two variables are related to each other. In the case of a linear equation, the coefficient of correlation is

$$r = \frac{n \, \Sigma xy - \Sigma x \, \Sigma y}{\sqrt{n \, \Sigma x^2 - (\Sigma x)^2} \; \sqrt{n \, \Sigma y^2 - (\Sigma y)^2}} \tag{40.4}$$

This coefficient is always between +1 and −1. A value of +1 or −1 would indicate that the two variables were totally correlated ; it would denote a perfect functional relationship, all the points

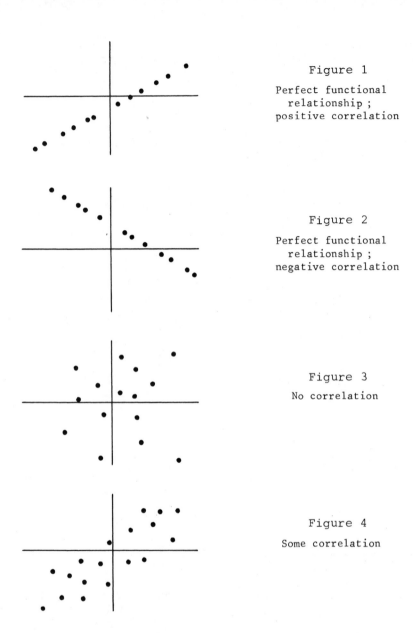

Figure 1

Perfect functional
relationship ;
positive correlation

Figure 2

Perfect functional
relationship ;
negative correlation

Figure 3

No correlation

Figure 4

Some correlation

representing paired values of x and y falling on the straight line representing this relationship. If $r = +1$, an increase of x corresponds to an increase of y (Figure 1). If $r = -1$, there is again a perfect functional relationship, but y decreases when x increases (Figure 2).

When $r = 0$, there is no relationship between x and y (Figure 3). In practice, however, when there is no relationship, one may find that r is not exactly zero, due to hazardous coincidences that generally occur except for an infinity of points.

When $|r|$ is between 0 and 1, there is a trend between x and y, although there is no strict relationship (Figure 4). Here, again, it should be noted that, *if* there is actually a strict relationship between the two variables, the calculation may give a value of r not exactly equal to +1 or -1, by reason of inaccuracies inherent to all measures.

Example 40.b : On page 10 of the Belgian journal *Heelal* of September 1978, the following table (Table 40.B) is given. For each of the twenty sunspot maxima which have occurred from 1761 to 1969, y is the height of the maximum (highest smoothed monthly mean), and x is the time interval, in months, since the previous sunspot minimum.

In this case, we find

$$\Sigma x = 1039 \qquad \Sigma x^2 = 57303 \qquad \Sigma xy = 108\ 987.0$$
$$\Sigma y = 2249.7 \qquad \Sigma y^2 = 286\ 027.09 \qquad n = 20$$

TABLE 40.B

epoch of maximum	y	x	epoch of maximum	y	x
1761 June	90.4	73	1870 July	144.8	39
1769 Oct.	125.3	38	1884 Jan.	78.1	61
1778 May	161.8	35	1893 Aug.	89.5	42
1787 Nov.	143.4	42	1905 Oct.	63.9	49
1804 Dec.	52.5	78	1917 Aug.	112.1	50
1816 March	50.8	68	1928 June	82.0	62
1829 June	71.5	74	1937 May	119.8	44
1837 Febr.	152.8	42	1947 July	161.2	39
1847 Nov.	131.3	52	1957 Nov.	208.4	43
1860 July	98.5	54	1969 Febr.	111.6	54

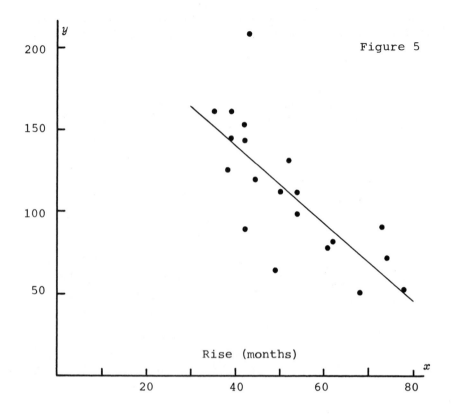

Figure 5

and then, by formulae (40.2) and (40.1),

$$y = 235.61 - 2.37\,x$$

This is the equation of the best straight line fitting the given twenty points. These points and the line are shown in Figure 5.

From formula (40.3) we find $r = -0.753$. This shows that there exists an evident trend to connexion, and the negative sign of r indicates that the correlation between x and y is negative : the *longer* the duration of the rise from a minimum to the next maximum of the sunspot activity, the *lower* this maximum generally is.

Example 40.c :   As an example of two variables between which we know there *cannot* exist any correlation, we consider the absolute photographic magnitude (x) of a minor planet and the number of the letters (y) in its name. We take the absolute magnitudes from the Leningrad *Ephemerides of Minor Planets* for 1979, and consider only the minor planets 51 to 100, which is a sufficient large sample (Table 40.C).

In this case, we have

$$\Sigma x = 460.8 \qquad \Sigma x^2 = 4278.50 \qquad \Sigma xy = 2869.1$$

$$\Sigma y = 312 \qquad \Sigma y^2 = 2078 \qquad n = 50$$

whence, by formula (40.3), $r = -0.097$. The coefficient of correlation is small, but not zero for the reason mentioned before.

TABLE 40.C

	x	y			x	y
51 Nemausa	8.7	7		76 Freia	9.0	5
52 Europa	7.6	6		77 Frigga	9.6	6
53 Kalypso	9.8	7		78 Diana	9.1	5
54 Alexandra	8.8	9		79 Eurynome	9.3	8
55 Pandora	9.1	7		80 Sappho	9.3	6
56 Melete	9.6	6		81 Terpsichore	9.7	11
57 Mnemosyne	8.4	9		82 Alkmene	9.4	7
58 Concordia	9.9	9		83 Beatrix	9.8	7
59 Elpis	8.7	5		84 Klio	10.3	4
60 Echo	10.0	4		85 Io	8.9	2
61 Danaë	8.8	5		86 Semele	9.8	6
62 Erato	9.8	5		87 Sylvia	8.3	6
63 Ausonia	8.2	7		88 Thisbe	8.2	6
64 Angelina	8.8	8		89 Julia	8.2	5
65 Cybele	7.9	6		90 Antiope	9.3	7
66 Maja	10.6	4		91 Aegina	9.7	6
67 Asia	9.9	4		92 Undina	8.0	6
68 Leto	8.3	4		93 Minerva	8.8	7
69 Hesperia	8.3	8		94 Aurora	8.8	6
70 Panopaea	9.2	8		95 Arethusa	8.9	8
71 Niobe	8.5	5		96 Aegle	9.1	5
72 Feronia	10.3	7		97 Klotho	8.7	6
73 Klytia	10.3	6		98 Ianthe	10.4	6
74 Galatea	10.1	7		99 Dike	11.5	4
75 Eurydike	10.0	8		100 Hekate	9.1	6

It should be noted that here, as in all other statistical studies, the sample must be sufficiently large in order to give a meaningful result. A correlation coefficient close to +1 or to −1 has no physical meaning if it is based on a too small number of cases. When we consider only the eight minor planets 71 to 78 in the example above, we find the high coefficient of correlation +0.785 between x and y, and even the still higher value $r = +0.932$ for the five minor planets 78 to 82.

This demonstrates that with too few cases the correlation coefficient can be accidently quite large.

As an exercice, show that there is no correlation between the rainfall at the Uccle Observatory and the sunspot activity, using the data of Table 40.D, where

y = total annual rainfall at Uccle, in millimeters,
x = yearly mean of the definitive sunspot numbers.

(Answer : the correlation coefficient is $r = -0.025$, which shows that there is no significant correlation between x and y.)

TABLE 40.D

year	y	x	year	y	x	year	y	x
1901	700	2.7	1928	882	77.8	1955	616	38.0
1902	762	5.0	1929	688	64.9	1956	795	141.7
1903	854	24.4	1930	953	35.7	1957	801	190.2
1904	663	42.0	1931	858	21.2	1958	834	184.8
1905	912	63.5	1932	858	11.1	1959	560	159.0
1906	821	53.8	1933	738	5.7	1960	962	112.3
1907	622	62.0	1934	707	8.7	1961	903	53.9
1908	678	48.5	1935	916	36.1	1962	862	37.5
1909	842	43.9	1936	763	79.7	1963	713	27.9
1910	990	18.6	1937	900	114.4	1964	785	10.2
1911	741	5.7	1938	711	109.6	1965	1073	15.1
1912	941	3.6	1939	928	88.8	1966	1054	47.0
1913	801	1.4	1940	837	67.8	1967	707	93.8
1914	877	9.6	1941	744	47.5	1968	776	105.9
1915	910	47.4	1942	841	30.6	1969	776	105.5
1916	1054	57.1	1943	738	16.3	1970	727	104.5
1917	851	103.9	1944	766	9.6	1971	691	66.6
1918	848	80.6	1945	745	33.2	1972	710	68.9
1919	980	63.6	1946	861	92.6	1973	690	38.0
1920	760	37.6	1947	640	151.6	1974	1039	34.5
1921	417	26.1	1948	792	136.3	1975	734	15.5
1922	938	14.2	1949	521	134.7	1976	541	12.6
1923	917	5.8	1950	951	83.9	1977	855	27.5
1924	849	16.7	1951	878	69.4	1978	767	92.5
1925	1075	44.3	1952	926	31.5	1979	839	155.4
1926	896	63.9	1953	557	13.9	1980	913	154.6
1927	837	69.0	1954	741	4.4	1981	1016	140.5

41

Ephemeris for Physical Observations of the Sun

The formulae given in this chapter are based on the elements determined by Carrington (1863), which have been in use since many years.
The required quantities are :

P = the position angle of the northern extremity of the axis of
rotation, measured eastwards from the North Point of the solar
disk ;

B_0 = the heliographic latitude of the central point of the solar
disk ;

L_0 = the heliographic longitude of the same point.

L_0 *decreases* by about 13.2 degrees per day. The mean synodic period is 27.2752 days. The beginning of each 'rotation' is the instant
at which L_0 passes through 0°. Rotation No. 1 commenced on 1853 November 9.

B_0 is zero about June 6 and December 7, and reaches maximum value
about March 6 (−7°.25) and September 8 (+7°.25).

Let JD be the Julian Ephemeris Date, which can be calculated by
means of the method described in Chapter 3. If the given instant is
in Universal Time, add to JD the value ΔT = ET − UT expressed in days
(see Chapter 5). If ΔT is expressed in seconds of time, the correction to JD will be + $\Delta T/86400$.

Then find T by means of formula (18.1), and calculate the following quantities :

$$\theta = (\text{JD} - 2398\,220) \times \frac{360°}{25.38}$$

$$I = 7°.25 = 7°15'$$

$$K = 74°.3646 + 1°.395\,833\,T$$

where I is the inclination of the solar equator on the ecliptic, and K is the longitude of the ascending node of the solar equator on the ecliptic.

Calculate the *apparent* longitude λ of the Sun (including the effect of aberration) by the method described in Chapter 18, and the obliquity of the ecliptic ε (including the effect of nutation) by means of formulae (18.4) and (18.5). Let λ' be λ corrected for the nutation.

Then calculate the angles x and y by means of

$$\tan x = - \cos \lambda' \tan \varepsilon$$

$$\tan y = - \cos (\lambda - K) \tan I$$

where both x and y should be taken between $-90°$ and $+90°$. Then the required quantities P, B_0 and L_0 are found as follows

$$P = x + y$$

$$\sin B_0 = \sin (\lambda - K) \sin I$$

$$\tan \eta = \frac{- \sin (\lambda - K) \cos I}{- \cos (\lambda - K)} = \tan (\lambda - K) \cos I$$

(η is in the same quadrant as $\lambda - K \pm 180°$)

$$L_0 = \eta - \theta, \quad \text{to be reduced to the interval } 0 - 360 \text{ degrees.}$$

Example 41.a : Calculate P, B_0 and L_0 for 1980 May 17 at 0^h UT = JD 2444 376.5.

We will use the value $\Delta T = +51^s = +0.00059$ day. Consequently, the corrected JD, or Julian Ephemeris Day, is JD = 2444 376.500 59. We then find successively

$$T = +0.803\ 737\ 183$$
$$\theta = 654\ 702°.1359 = 222°.1359$$
$$I = 7°.25$$
$$K = 75°.4865$$

From Chapter 18 :

$$L = 29\ 214°.85346 = 54°.85346$$
$$M = 29\ 292°.25055 = 132°.25055$$
$$C = +1°.39820$$
$$\odot = 56°.25166$$

$$\Omega = -1295°.36 = +144°.64$$

Correcting for aberration only :

$$\lambda = \Theta - 0°.00569 = 56°.24597$$

Correcting for nutation :

$$\lambda' = \lambda - 0°.00479 \sin \Omega = 56°.24320$$

$$\varepsilon = 23°.441\,835$$

$\tan x = -0.240\,941$	$x = -13°.5467$
$\tan y = -0.120\,110$	$y = -6°.8490$

$$P = -20°.40$$

$$\sin B_0 = -0.041\,587 \qquad B_0 = -2°.38$$

$$\tan \eta = \frac{+0.326\,900}{-0.944\,143} \qquad \eta = 160°.9021$$

$$L_0 = -61°.2338 = 298°.77$$

The $A.E.$ for 1980, page 355, gives

$$P = -20°.40 \qquad B_0 = -2°.38 \qquad L_0 = 298°.76$$

As mentioned above, a solar "rotation" begins when $L_0 = 0°$. An approximate time for the beginning of Carrington's synodic rotation No. C is given by

$$\text{Julian Date} = 2398\,140.24 + 27.275\,232\,C$$

where, of course, C is an integer. The instant so obtained will not be in error by more than 0.15 day approximately.

For example, for $C = 1699$, one finds $JD = 2444\,480.86$, which corresponds to 1980 August 29.36. The correct value, according to the $A.E.$, is August 29.22.

A correct value for the time can be obtained by calculating L_0 for two instants near the time given by the formula above, and then by performing an inverse interpolation to find when L_0 is zero.

42

RISING, TRANSIT AND SETTING

Although formulae concerning the rising and setting of celestial bodies have already been given in Chapter 8, the practical method described in the present chapter might be of interest.

We will use the following symbols :

L = longitude of the observer in degrees, measured *positively west
from Greenwich*, negatively to the east ;

ϕ = latitude of the observer, positive in the northern hemisphere,
negative in the southern hemisphere ;

ΔT = the difference ET – UT in *seconds* of time ;

h_0 = the "standard" altitude, *i.e.* the geometric altitude of the center of the body at the time of apparent rising or setting,
namely

h_0 = $-0°34'$ = $-0°.5667$ for stars and planets ;
$-0°50'$ = $-0°.8333$ for the Sun.

For the Moon, the problem is more complicated because h_0 is not
constant. Taking into account the variations of semidiameter
and parallax, we have for the Moon

$$h_0 = 0.7275\,\pi - 0°34'$$

where π is the Moon's horizontal parallax. If no great accuracy is required, the mean value $h_0 = +0°.125$ can be used for
the Moon.

Suppose we wish to calculate the times, in *Universal Time*, of rising, of transit (when the body crosses the local meridian at upper
culmination) and of setting of a celestial body at the observer's
place on a given date D. We take the following values from an almanac, or we calculate them ourselves with a computer program :

- the apparent sidereal time θ_0 at 0^h *Universal Time* on day D for the Greenwich meridian, converted into *degrees* ;
- the apparent right ascensions and declinations of the body

α_1 and δ_1 on day $D-1$ at 0^h Ephemeris Time
α_2 and δ_2 on day D —
α_3 and δ_3 on day $D+1$ —

The right ascensions should be expressed in *degrees* too.

We first calculate *approximate* times as follows.

$$\cos H_0 = \frac{\sin h_0 - \sin \phi \sin \delta_2}{\cos \phi \cos \delta_2} \qquad (42.1)$$

Express H_0 in degrees. H_0 should be taken between $0°$ and $+180°$. Then we have :

for the transit : $m_0 = \dfrac{\alpha_2 + L - \theta_0}{360}$

for the rising : $m_1 = m_0 - \dfrac{H_0}{360}$

for the setting : $m_2 = m_0 + \dfrac{H_0}{360}$

These three values m are times, on day D, expressed as fractions of a day. Hence, they should be between 0 and +1. If one or more of them are outside this range, add or subtract 1. For instance, +0.3744 should remain unchanged, but −0.1709 should be changed to +0.8291, and +1.1853 should be changed to +0.1853.

Now, for *each* of the three m-values *separately*, perform the following calculation.

Find the sidereal time at Greenwich, in *degrees*, from

$$\theta = \theta_0 + 360.985\,647\,m$$

where m is either m_0, m_1 or m_2.

For $n = m + \Delta T/86400$, interpolate α from α_1, α_2, α_3 and δ from δ_1, δ_2, δ_3, using the interpolation formula (2.3). For the calculation of the time of transit, δ is not needed.

Find the local hour angle of the body from $H = \theta - L - \alpha$

and then the body's altitude by means of formula (8.6). This altitude is not needed for the calculation of the time of transit.

Then the correction to m will be found as follows :

- in the case of a transit,

$$\Delta m = -\frac{H}{360}$$

where H is expressed in degrees and *must* be between −180 and +180 degrees. (In most cases, H will be a small angle and be between −1° and +1°) ;

- in the case of a rising or setting,

$$\Delta m = \frac{h - h_o}{360 \cos \delta \cos \phi \sin H}$$

where h and h_o are expressed in degrees.

The corrections Δm are small quantities, in most cases being between −0.01 and +0.01.

The corrected value of m is then $m + \Delta m$. If necessary, a new calculation should be performed using the new value of m.

At the end of the calculation, each value m should be converted into hours by multiplication by 24.

Example 42.a : Venus on 1980 March 26 at Boston,
longitude = +71°05′ = +71°.0833 ,
latitude = +42°20′ = +42°.3333.

We take from the *Astronomical Ephemeris* :

1980 March 26, 0^h UT : $\theta_o = 12^h14^m21^s.882$ = 183°.59118

1980 March			
25.0 ET :	$\alpha_1 =$	$3^h07^m03^s.72$	= 46°.76550
26.0	$\alpha_2 =$	3 11 17.47	= 47.82279
27.0	$\alpha_3 =$	3 15 30.98	= 48.87908

1980 March			
25.0 ET :	$\delta_1 =$	+20°08′01″.5	= +20°.13375
26.0	$\delta_2 =$	+20 29 14.3	= +20.48731
27.0	$\delta_3 =$	+20 49 59.6	= +20.83322

We take $h_o = -0°.5667$, $\Delta T = +51^s$, and find by formula (42.1) $\cos H_o = -0.354\,659$, $H_o = 110°.7726$, whence the approximate values :

$$\begin{array}{lll}
\text{transit :} & m_o = -0.17968, & \text{whence } m_o = +0.82032 \\
\text{rising :} & m_1 = m_o - 0.30770 = +0.51262 \\
\text{setting :} & m_2 = m_o + 0.30770 = +1.12802, & \text{whence } +0.12802
\end{array}$$

Calculation of more exact values :

		rising	*transit*	*setting*
	m	+0.51 262	+0.82 032	+0.12 802
	θ	8°.63 964	119°.71 493	229°.80 456
	n	+0.51 321	+0.82 091	+0.12 861
inter-	α	48°.36 501	48°.68 998	47°.95 870
polation	δ	+20°.66 579		+20°.53 223
	H	−110°.8087	−0°.0584	+110°.7626
	h	−0°.46 109		−0°.52 779
	Δm	−0.000 45	+0.000 16	+0.000 17
corrected	m	+0.51 217	+0.82 048	+0.12 819

A new calculation, using these new values of m, yields the new corrections −0.000 004, +0.000 002 and −0.000 003, respectively, which thus can be neglected. We then have, finally :

$$\begin{array}{llll}
\text{rising :} & m_1 = +0.51\,217, & 24^h \times 0.51217 = 12^h18^m \text{ UT} \\
\text{transit :} & m_o = +0.82\,048, & 24^h \times 0.82048 = 19^h41^m \text{ UT} \\
\text{setting :} & m_2 = +0.12\,819, & 24^h \times 0.12819 = 3^h05^m \text{ UT}
\end{array}$$

NOTES

1. In Example 42.a we found that at Boston the time of setting was 3^h05^m UT on March 26. However, converted to *local* standard time this corresponds to an instant on the evening of the previous day ! If really the time of setting on March 26 is needed in local time, the calculation should be performed using the value $m_2 = +1.12802$ first found, instead of +0.12802.

2. If the body is circumpolar, the second member of formula (42.1) will be larger than 1 in absolute value, and there will be no angle H_o. In such a case, the body will remain the whole day either above or below the horizon.

43

Heliocentric Position of Pluto

As mentioned at the end of Chapter 24, no theory for the motion of Pluto is available. However, expressions for an accurate representation of the planet's motion for the years 1885 to 2099 have been constructed by E. Goffin, J. Meeus and C. Steyaert. They have been published in *Astronomy and Astrophysics*, Vol. 155, pages 323-325 (1986), and are reprinted here, except that the smallest periodic terms have been dropped.

Let JD be the Julian Ephemeris Day corresponding to the given instant. Then find

$$T = \frac{JD - 2415\,020.0}{36525}$$

and calculate the following angles (in degrees):

$$J = 238.74 + 3034.9057\,T$$
$$S = 267.26 + 1222.1138\,T$$
$$P = 93.48 + 144.9600\,T$$

Then the heliocentric longitude l and latitude b, expressed in degrees and referred to the standard equinox of 1950.0, and the radius vector r in astronomical units, are

$$l = 93.297\,471 + 144.9600\,T + \Sigma l$$
$$b = -3.909\,434 + \Sigma b$$
$$r = 40.724\,725 + \Sigma r$$

where the sums of periodic terms, Σl, Σb and Σr, are given on the next page. Calculated in this way, l and b will be less than 1" in error, and the radius vector will be less than 0.0001 a.u. in error, with respect to the numerical integration on which this representation of Pluto's motion is based. It's important to note, as has been said, that the method given here is not valid outside the years 1885 to 2099.

Σl : periodic terms in longitude (in units of the sixth decimal of a degree)

-19977972 sin P	-738 sin $(S+3P)$	+394 sin $(J+P)$
+19667536 cos P	+3443 cos $(S+3P)$	-55 cos $(J+P)$
+987114 sin $2P$	+1234 sin $(2S-2P)$	+119 sin $(J+2P)$
-4939350 cos $2P$	+472 cos $(2S-2P)$	-264 cos $(J+2P)$
+577978 sin $3P$	+1101 sin $(2S-P)$	-46 sin $(J+3P)$
+1226898 cos $3P$	-894 cos $(2S-P)$	-156 cos $(J+3P)$
-334695 sin $4P$	+625 sin $2S$	-77 sin $(J+4P)$
-201966 cos $4P$	-1214 cos $2S$	-33 cos $(J+4P)$
+130519 sin $5P$	+2485 sin $(J-S)$ ·	-34 sin $(J+S-3P)$
-29025 cos $5P$	-486 cos $(J-S)$	-26 cos $(J+S-3P)$
-39851 sin $6P$	+852 sin $(J-S+P)$	-43 sin $(J+S-2P)$
+28968 cos $6P$	-1407 cos $(J-S+P)$	-15 sin $(J+S-P)$
+20387 sin $(S-P)$	-948 sin $(J-3P)$	+21 cos $(J+S-P)$
-9832 cos $(S-P)$	+1073 cos $(J-3P)$	+10 sin $(2J-3P)$
-3986 sin S	-2309 sin $(J-2P)$	+22 cos $(2J-3P)$
-4954 cos S	-1024 cos $(J-2P)$	-57 sin $(2J-2P)$
-5817 sin $(S+P)$	+7047 sin $(J-P)$	-32 cos $(2J-2P)$
-3365 cos $(S+P)$	+770 cos $(J-P)$	+158 sin $(2J-P)$
-3903 sin $(S+2P)$	+1184 sin J	-43 cos $(2J-P)$
+2895 cos $(S+2P)$	-344 cos J	

Σb : periodic terms in latitude (in units of the sixth decimal of a degree)

-5323113 sin P	+312 sin S	-177 sin $(J-S)$
-15024245 cos P	-128 cos S	+259 cos $(J-S)$
+3497557 sin $2P$	+2057 sin $(S+P)$	+15 sin $(J-S+P)$
+1735457 cos $2P$	-904 cos $(S+P)$	+235 cos $(J-S+P)$
-1059559 sin $3P$	+19 sin $(S+2P)$	+578 sin $(J-3P)$
+299464 cos $3P$	-674 cos $(S+2P)$	-293 cos $(J-3P)$
+189102 sin $4P$	-307 sin $(S+3P)$	-294 sin $(J-2P)$
-285383 cos $4P$	-576 cos $(S+3P)$	+694 cos $(J-2P)$
+14231 sin $5P$	-65 sin $(2S-2P)$	+156 sin $(J-P)$
+101218 cos $5P$	+39 cos $(2S-2P)$	+201 cos $(J-P)$
-29164 sin $6P$	-97 sin $(2S-P)$	+294 sin J
-27461 cos $6P$	+208 cos $(2S-P)$	+829 cos J
+4935 sin $(S-P)$	-160 cos $2S$	-123 sin $(J+P)$
+11282 cos $(S-P)$		-31 cos $(J+P)$

Σr : periodic terms in radius vector (in units of 0.000 001 a.u.)

+6623876 sin P	−3 sin $(S+2P)$	−4 sin $(J-P)$
+6955990 cos P	+79 cos $(S+2P)$	+4564 cos $(J-P)$
−1181808 sin $2P$	+50 sin $(S+3P)$	+852 sin J
−54836 cos $2P$	+54 cos $(S+3P)$	+855 cos J
+163227 sin $3P$	− sin $(2S-2P)$	−88 sin $(J+P)$
−139603 cos $3P$	−22 cos $(2S-2P)$	−82 cos $(J+P)$
−3644 sin $4P$	+84 sin $(2S-P)$	+21 sin $(J+2P)$
+48144 cos $4P$	−48 cos $(2S-P)$	−12 cos $(J+2P)$
−6268 sin $5P$	−30 sin $2S$	−14 sin $(J+3P)$
−8851 cos $5P$	+61 cos $2S$	+6 cos $(J+3P)$
+3111 sin $6P$	+26 sin $(J-S)$	−6 sin $(2J-3P)$
−408 cos $6P$	−39 cos $(J-S)$	+ cos $(2J-3P)$
−621 sin $(S-P)$	−19 sin $(J-S+P)$	+13 sin $(2J-2P)$
+2223 cos $(S-P)$	−40 cos $(J-S+P)$	−23 cos $(2J-2P)$
+438 sin S	−321 sin $(J-3P)$	+25 sin $(2J-P)$
+450 cos S	+42 cos $(J-3P)$	+107 cos $(2J-P)$
−153 sin $(S+P)$	+797 sin $(J-2P)$	+25 sin $2J$
+61 cos $(S+P)$	−792 cos $(J-2P)$	+16 cos $2J$

Example 43.a : For 1977 January 7.0 ET, we find

$$JD = 2443\,150.5 \qquad T = 0.770\,171\,1157$$

$J = 2576°.1367 = 56°.1367 \qquad \Sigma l = -13\,084\,570$
$S = 1208°.4967 = 128°.4967 \qquad \Sigma b = +20\,904\,572$
$P = 205°.1240 \qquad\qquad\quad\; \Sigma r = -10\,189\,234$

from which

$l = 204°.941\,476 - 13°.084\,570 = 191°.8569$

$b = -3°.909\,434 + 20°.904\,572 = +16°.9951$

$r = 40.724\,725 - 10.189\,234 = 30.5355$ a.u.

The calculation of the geocentric positions can be obtained from the heliocentric ones, by means of the usual methods.

INDEX

The numbers refer to the pages